Produsing Theory
in a Digital World

Steve Jones
General Editor

Vol. 80

The Digital Formations series is part of the Peter Lang Media and Communication list.
Every volume is peer reviewed and meets
the highest quality standards for content and production.

PETER LANG
New York • Washington, D.C./Baltimore • Bern
Frankfurt • Berlin • Brussels • Vienna • Oxford

Produsing Theory in a Digital World

The Intersection of Audiences and Production in Contemporary Theory

Edited by
Rebecca Ann Lind

PETER LANG
New York • Washington, D.C./Baltimore • Bern
Frankfurt • Berlin • Brussels • Vienna • Oxford

Library of Congress Cataloging-in-Publication Data

Produsing theory in a digital world: the intersection of audiences and production
in contemporary theory / edited by Rebecca Ann Lind.
p. cm. — (Digital formations; v. 80)
Includes bibliographical references and index.
1. Mass media—Technological innovations. 2. Mass media and technology.
3. Mass media—Social aspects. 4. Media literacy. 5. Digital media. 6. Social
media. I. Lind, Rebecca Ann. II. Title: Produsing theory in a digital world.
P96.T42P76   302.23—dc23   2012014097
ISBN 978-1-4331-1520-2 (hardcover)
ISBN 978-1-4331-1519-6 (paperback)
ISBN 978-1-4539-0840-2 (e-book)
ISSN 1526-3169

Bibliographic information published by **Die Deutsche Nationalbibliothek.**
Die Deutsche Nationalbibliothek lists this publication in the "Deutsche
Nationalbibliografie"; detailed bibliographic data is available
on the Internet at http://dnb.d-nb.de/.

Cover art image AGENCY by Chuck Sabec (2012).
Reprinted with kind permission of the artist and Studio Eight Fine Art
(www.studioeightfineart.com). All rights reserved.

Layout of interior text by ChiTownMuggle.

Contents

Acknowledgments

My greatest thanks are to the contributors to this volume. It has been a pleasure working with them; besides bringing a plethora of intriguing ideas to these pages they have been responsive, understanding, and willing to engage in some fairly intense conversations with me during the writing process. I hope they are pleased with the outcome.

As always, many thanks are due to my colleagues (especially Steve Jones and Zizi Papacharissi) in the Department of Communication at the University of Illinois at Chicago, and to the College of Liberal Arts and Sciences, for their support.

Grateful acknowledgment is made to the following for permission to use copyrighted material:

Cover image AGENCY by Chuck Sabec (2012). Reprinted with kind permission of the artist and Studio Eight Fine Art (www.StudioEightFineArt.com). All rights reserved.

Figures 11.1 and 11.2 adapted from Webb, L. M., & Lee, B. S. (2011). Mommy blogs: The centrality of community in the performance of online maternity. In M. Moravec, Ed., *Motherhood online: How online communities shape modern motherhood* (pp. 244-257). Newcastle upon Tyne, England: Cambridge Scholars Publishing. Published with permission of Cambridge Scholars Publishing.

To my parents and my students.

CHAPTER ONE

Produsing Theory in a Digital World: Illuminating *Homo Irretitus*

Rebecca Ann Lind

In…many…spaces…, users are always already necessarily also producers of the shared knowledge base, regardless of whether they are aware of this role—they have become a new, hybrid, *produser*. (Bruns, 2008, p. 2, emphasis in original)

Are we all produsers now? (Bird, 2011, p. 502)

The timeline design that Facebook began to roll out in 2011 was, like all Facebook changes, met with both praise and condemnation. I enjoyed the timeline's bolder design—in part because of the combination of the profile and cover images—so I made the switch fairly early. It didn't take long before I was reminded that our creativity on Facebook is bounded. On the one hand, Facebook proclaims: "Your cover appears above your profile picture, and it's the first thing people see when they visit your timeline. Choose any image you want and change it as often as you like." Yet on the other hand, if your desired image is outside the parameters, you'll be told to "Try a different image. Please choose an image that's at least 399 pixels wide." So much for "choos[ing] any image you want." Properly admonished for selecting an insufficiently wide image, I quickly found an appropriate replacement. And in the process I learned to censor myself, and to filter out future possible cover images that wouldn't meet the criteria.

As users of what are frequently referred to as "new media," "social media," "emerging media" and "Web 2.0"—all labels that are at least somewhat problematic—we constantly navigate the space between freedom and control. Simultaneously, we're also navigating spaces of self, and of other: creating, recreating, and performing our identities and our relations with others. Obviously, we're also navigating the space between the production and use of media.

This book is about some of those navigations and some of those tensions; it is an intersection of theories layered onto an intersection of previously separate elements of the mass communication system. The Audience—Content—Production trichotomy is still a convenient classification system in some settings (e.g., Lind, 2013), but we must also attend to the intersections of these elements in order to understand the evolving nature of our contemporary media environment. Indeed, each of these elements, as discrete components, has long been under attack by theorists. Years before Rosen (2006) referred to "the people formerly known as the audience," Fiske (1989) questioned the viability not only of not only the television audience but also of television's texts: "There is no text, there is no audience, there are only the processes of viewing,—that variety of cultural activities that take place in front of the screen" (p. 57). Previously, Barthes (1967/1977) had relieved the author of any claim to omnipotence, going so far as to declare "the death of the author" (p. 142) in his quest to acknowledge the primacy of (audience) interpretations of any text. Finally, despite the position of screen theory adherents in film studies (e.g., Heath, 1977; MacCabe, 1974), most communication scholars never endowed the text with extraordinary powers to determine meaning—even before it (or at least a subset thereof) was reconceptualized into what Bruns (2008) presented as a mutable *artifact* rather than a fixed product. It's not that these components don't exist or aren't valuable; rather that they need to be reconsidered in their current intersectional context. We may still refer to texts, audiences, and producers, but fully understanding them in the 21st century, especially in a media environment well populated with, if not yet saturated by, social software (defined by Coates [2005] as "software which supports, extends, or derives added value from, human social behaviour—message-boards, musical taste-sharing, photo-sharing, instant messaging, mailing lists, social networking"[1]) means exploring how these components interact.

This volume gathers together scholars from not only from within but also external to the core of new media studies. Each of the authors here considers some component of the intersection of audience and production in a terrain enabled by emerging communications technologies. Each chapter is concerned with some form of social software, considering a subset of the phenomenon for which Axel Bruns (2008) has provided an effective label: *produsage*, briefly defined as "the collaborative and continuous building and extending of existing content in pursuit of further improvement" (Bruns, 2008, p. 21).

The intersectional spaces considered in this book, enabled by social software, are populated by the entity Saulauskas (2000) called "*Homo Irreti-*

tus" ("irretitus" meaning "caught in a net, trapped"), and described as "one who nets in the net: the netting and at the same time netted human being." Saulauskas is not alone in concluding that both the hopes (dreams) and the fears (superstitions) of *Homo Irretitus* at the turn of the century were unlikely to be realized. The Internet *can* spawn utopia or dystopia; thus far, it has produced (we might even say "prodused") neither. The Internet is a technology, a tool; we will get out of it what we make of it, through our social practices and engagement with, in, and through the spaces and contexts it facilitates.

But the very making of our social practices, spaces, and contexts in this brave new world[2] of the World Wide Web, the work of *Homo Irretitus* in this intersectional space, must be interrogated. If we are to understand this space, we should approach it from varied vantage points. The multifaceted, converging perspectives of the theoretical approaches used in this volume help shed light onto a variety of the tensions evident in the new digital spaces in which we create and re-create (and often produse) so much of our lives, our identities, our selves. The space may be new, but because it has not generated a utopia, many of the traditional tensions or concerns of our social world—such as structurally induced limitations on human agency—remain. This is of course no surprise; no matter how flexible it may be, the World Wide Web cannot be expected to subvert all of the extant social structures such as family, church, education, religion, government, and so forth.

Many of the chapters in this volume, even though the authors have been guided by disparate theories (and even though such considerations are not necessarily the primary themes of their contributions), address tensions between structure and agency. Thus, it is worthwhile to turn our attention to this tension here. This theme is a hardy perennial in our field, having long intrigued a variety of theorists. The most helpful perspectives for our purposes are those that acknowledge the mutual influence of social structures and human interactions, such as the influential work by Berger and Luckmann (1967) on the social constructivist paradigm. Still, the nature of the relationship between structure and agency has been conceptualized in multiple ways. Bourdieu (1979/1984), also positing a dialectical relationship between structure and agency, added to the conversation the helpful concepts of habitus, field, and capital. Giddens' (1984) structuration theory, frequently referenced in communication studies, considers both structure and agency to operate simultaneously, and adds the useful concept of "reflexivity," in which agents hold the capability of acting in such a way as to effect change (or, conversely, to perpetuate the status quo). Thus, and importantly, emancipatory potential is ever-present. Unger (2001), in his anti-necessitarian social theory, took the conceptualizations of agency and empowerment further, making them more flexible, more varied, and hence more attainable. In so doing, he invoked what he called "the poet's turn of phrase," using the somewhat ungainly term "negative capability" to represent the possibility of "empowerment that arises from the denial of whatever in our contexts delivers us over to a fixed scheme of division and hierarchy and to an enforced choice between routine and rebellion" (p. 279). Unger argued that social institutions can be "denaturalized" (pp. 125–126), and that individuals may seek negative capability with varying degrees of intentionality, which he related to agency. There is therefore a spectrum of possibility for behaviors representing negative capability, ranging across and through intentional agency, non-intentional agency, and partly deliberative agency (pp. 294–302). Unger's conceptualization is useful in that it is not a binary consideration but rather offers a nuanced view of agency as well as of different forms of empowerment (e.g., economic and non-economic).

A still more refined, and fuller, viewpoint is provided in Dreier's (2008) theory of persons as situated participants in social practices. Dreier's approach, rooted in critical psychology and social practice theory, is intriguing and seems particularly well suited for an exploration of agency and structure at the intersection of audience and production. He argued that a Habermasian conceptualization of social practice as an opposition between the system and the everyday life world (or indeed, any other of the dualities presented in theorizing about structure and agency) is less useful than "a conception of a structure of ongoing social practice in a set of linked and diverse, local social contexts" (p. 23). Dreier's theory holds that social practices produce and reproduce the social world; and that social structures and social practices are interrelated and mutually influencing, with participants—diverse, changing, connected, ever-acting—the interface between the two. Dreier argued that:

> Notions about an abstract, individual agency must be replaced by a contextual conception of personal modes of participation rendering personal abilities many-sided and variable. We may then ask how personal stability and change are allowed and inhibited by a person's trajectory of participation in structures of social practice. (p. 40)

To Dreier, individuals act and interact within and across a variety of *social contexts*, defined as "a delineated, local place in social practice that is reproduced and changed by the linked activities of its participants and through its links with other places in a structure of social practice" (p. 23). The goodness-of-fit between this definition, an intersection made possible through the use of social software, and the concept of produsage is immediately apparent. Social contexts may be fleeting or long term; they may be more or less open to participants; they may to greater or lesser extent limit the range of possi-bilities in terms of the social practices to be carried out within them; their structural arrangements may be more or less definitive; and importantly, they can only be understood in the larger structure of social practice, and in their interconnections with or separations from other contexts.

Dreier's inclusion of the *local* in his definitions is an effort to re-introduce the situated nature of activity; he challenged theorists such as Gid-dens, who claimed that "[p]lace becomes phantasmagoric," because it is "thoroughly penetrated by disembedding mechanisms, which recombine the local activities into time-space relations of ever-widening scope" (Giddens, 1991, p. 146). Should disembedding or detachment imply that place is irrele-vant? Not to Dreier, who argued that "the problem" is that Giddens and other "influential theorists confuse being situated with being situation-bound," and that even in an increasingly globalized world, "personal social practices are not global. They keep on being situated in and across particular locations, that is, translocal, no matter how scattered the locations…" (Dreier, 2008, p. 27).

To Dreier, theorizing persons as *participants* in a social practice is key for a number of reasons, not the least of which is that doing so brings to-gether all of the elements that otherwise would remain ungrounded, and that it highlights the importance of "what [persons] are part of and how they in-volve themselves in it" (p. 30). Dreier also noted that his focus on participa-tion acknowledged that as parts in and of various social practices, persons play different roles in different contexts; that each role played in each con-text represents only a part of the social practice; and that through their par-ticipation, persons may work to reproduce or alter the practice. Participants hold varying social *positions* and *locations*; locations are "quasi-physical" and socio-material, whereas positions are more closely related to the "struc-tured, institutionalized, and contextualized social landscape" (p. 32).

As people participate in varied social contexts, personal *stances* may emerge. Dreier emphasized that stances are not standpoints,[3] which are more firmly rooted in social positions; rather, stances can be quite malleable, and are generated by persons participating in multiple contexts as they "evaluate, link, and generalize premises" (p. 42) across contexts. Persons adopt multiple stances, depending on the links and the premises being drawn upon in the re-levant contexts. Importantly, the concept of stance acknowledges the variety and diversity among persons sharing any given social position, or even any given intersection of social positions, thereby lessening the extent to which we see such social positions as determinative in and of themselves.

Dreier's theory of persons as situated participants in social practice has much to offer scholars in mass communication interested in agency, empowerment, identity, and other considerations related to social interaction, and of course to scholars interested in personality and psychotherapy. Focusing on the intersection of the social practices of situated persons, while acknowledging the breadth and range of their local and translocal social contexts and the variability of their stances, as well as the role of their social positions, properly highlights the richness of the tensions presented to actors as they engage in and with the social world, and as, through their interactions, they work both to reproduce and to challenge social structures. Doing so may enrich a variety of research efforts related to the incorporation of social software into our daily lives, whether we are using that software to produse or more simply to engage, or to cache our life events. Perhaps, as we remediate theory (see Papacharissi, this volume), we may temporarily revise—or at least revisit—our stances as scholars to see whether we wish to complicate our considerations of the subject at hand.

Of course, there are many other concepts that can and do enrich our understanding of the intersection of media production and consumption, as an overview of the contributions to this volume will attest. Following Papacharissi's advice to play with linearity, rather than proceed in a numerically sequential fashion through the chapters, let us stay a while longer with our primary theme (agency/empowerment), before turning to others.

Jay Bolter (Chapter 3) provides a good starting point for our more situated discussion of the tension between freedom and control, in that he reinforces the extent to which we, as the ones immersed in the interactive digital environment, are doing much more than merely playing or using tools of work or play. We are learning, we are being trained—in this case, in how to be procedural. Bolter explores the concept of procedurality, an outgrowth of industrial-age mechanization. He highlights two key elements of procedurality: parameterization and event loops. He argues that procedurality is becoming an increasingly important technical and cultural response in this era of digital communication and entertainment, that many forms of digital media explore and celebrate procedurality, and that videogames teach their players how to function in a procedural world. Rituals of procedurality are prescribed through various forms of social media used by people to perform and communicate their identities; they are dictated not only by cultural agreements but also by the code underlying the technology. The procedures are both limiting and flexible. However, questions remain about the human impact of this new form of mechanization, despite people's willingness to perform in and through such digital procedures.

Shayla Thiel-Stern (Chapter 6) writes about the agency-structure tension both in terms of technological and societal concerns. She theorizes the concept of identity using Goffman's and Butler's concepts of the public performance of identity as a starting point. Her consideration of adolescents' use of social media is informed by and extends Goffman's dramaturgical conceptualizations of front stage and back stage. She argues that young people's public identity production is iterative and collaborative, and that it thrives in the very space of the theatrical fourth wall. In this context, the fourth wall does not separate the performer from the audience; instead, it provides a robust environment in which young people present and represent themselves. Thiel-Stern further complicates public identity production in this context by reflecting on the increasingly templated (i.e., parameterized) nature of social networking tools and by acknowledging that this public negotiation of identity still takes place in the broader cultural discourse regarding normative behavior, and in which the dominant ideologies continue to resonate. Young people can articulate their identity on their own terms, but not with absolute agency. Thiel-Stern argues that teens are in some ways empowered by the new communications media, and that they do engage in produsage activities, but much of the material they create reinforces cultural stereotypes, such as those about sex and gender. Even though young people (and others) are clearly using interactive media to enact and perform identity, they are doing so within a cultural landscape containing numerous social structures that perpetuate the dominant ideology.

Guided by queer theory, Gust Yep, Miranda Olzman, and Allen Conkle (Chapter 8) also found evidence of hegemonic influence, in this case in videos uploaded by queers. Their chapter is a report on their analysis of video messages posted on the It Gets Better project, a site designed to offer hope and encouragement to members of LGBTQ[4] communities. Focusing on videos posted by non-mainstream LGBTQ individuals, the authors consider how this user-generated content reflects progress narratives, the politics of affect, and queer world-making. In many ways, the stories analyzed do not challenge heteronormativity as fully as possible: often, they focus on individual or spiritual concerns rather than the larger sociocultural milieu, normalize violence and downplay ugly feelings, and adhere to a politic of assimilation and normativity. The very premise of the It Gets Better Web site, which arguably *should* be questioned, may not reflect the reality faced by many who don't fit into cultural norms.

Diego Costa's (Chapter 9) consideration of user-generated content on cuckold porn sites further demonstrates that the racism at work in our material world is reflected in the virtual world. Costa explores the world of amateur pornography to see how user-generated content (videos and comments) depicting the cuckold fantasy functions on multiple levels. Focusing on the interracial cuckold fantasy, guided primarily by post-colonial theory, he argues that such content represents a colonial encounter, that the sexual acts involved reflect a virtual outsourcing of labor, that such labor is perverse labor, and that the black penis—which stands in stark contrast to the white phallus—functions as a prosthesis. As he considers these points, Costa interrogates sexuality and difference, as well as raced and gendered politics and power.

Paul Booth (Chapter 5) brings his concept of demediation to the interactive role playing adventure game, *MagiQuest*. Presented as a position between hypermediation and immediacy, demediation represents the space where we simultaneously look at and look through media. In *MagiQuest,* technology is fully incorporated into the players' production of the game; it is a practice more than a product. Booth argues that the mediation is so prevalent in the game that players are taught to use technology instrumentally. Ultimately, players are taught to be passive consumers of media information even as they are taught to be active producers of information and play. Their imaginative and active engagement with the game coexists with the reproduction of the dominant ideology. Playspaces are also learning spaces, and Booth draws upon Althusser's concept of interpellation, in which a subject is prescribed a set of characteristics. Through their play, players implicitly and explicitly reinforce characteristics related to traditional gender norms, whiteness as normative, consumption, and segmentation. Booth argues that the game is a metaphor for life and a guide for how to live—a constituted, capitalistic life.

It seems clear that members not only of the dominant cultural group, but also those in subordinated groups, are producing, reproducing, and produsing hegemonic cultural barriers. That the dominant groups do so (as noted by Booth) is less surprising than that the queers (Yep, Olzman, & Conkle), people of color (Yep, Olzman, & Conkle; to a limited extent Costa), and young people (Thiel-Stern) do so.

Of course we can't expect interactive media themselves, by themselves, to eliminate racism, classism, sexism, and so forth (despite Lee and Webb's [Chapter 11] reminder that mommy blogging can indeed be a radical act), but that gets even harder when, as Booth and Thiel-Stern note, we are so obviously being interpellated, or hailed, as we actively immerse ourselves into the process (the event loops)—even when we do acknowledge that we are not simply determined by the structures hailing us. As we play, we generate the content, and create a (virtual) reality, and in so doing, learn. As previously mentioned, both Bolter and Booth overtly consider the pedagogical power of play, and there is clearly a playful aspect of some implementations of the technobiographic technologies Freedman (discussed below) considers.

The challenge of pushing back against the processes of interpellation becomes that much greater when so much of the environment within which we are playing, produsing, or working is controlled by large corporations. Booth and Bolter both reference the commercial environment, an environment discussed more fully by Freedman.

Eric Freedman's (Chapter 4) considerations of the life technobiographic (life reflected by a data trail, written through technology), and how subjectivity is transformed by new communication technologies, illustrate subtle dimensions of political economy. Freedman argues that studying material devices (e.g., the iPhone) will only provide a partial view; we must also attend to the immaterial frameworks (e.g., the software, networks, and so forth) undergirding the devices. His chapter provides an overview of the industrial roots of technologies that are at once tools of communication and technobiographic agents, discusses the networked body, and highlights the tension between the presumed agency afforded by these tools and the limitations imposed by the commercial paradigm in which they are generated.

Importantly, because social media are heavily proceduralized, the freedoms associated with such activities are limited. Thiel-Stern's consideration of agency in the use of social media, raised in her discussion of the templated nature of such media, is linked directly to each medium's parameters. Although social media can and do allow users to negotiate their identities, one cannot—as noted previously—upload material that is too big, too small, not of a proper format, and so forth.

Still, engaging in produsage has great potential to reduce the power of social structures, and could even reshape democracy. As Bruns put it:

> What may result from this renaissance of information, knowledge, and creative work, collaboratively developed, compiled, and shared under a produsage model, may be a fundamental reconfiguration of our cultural and intellectual life, and thus of society and democracy itself. (2008, p. 34)

Two chapters in this collection most closely reflect the classic application of produsage in an information- or knowledge-oriented context. Bruns and Highfield turn to citizen journalism in Twitter; Hills investigates fan spoilers.

Axel Bruns and Tim Highfield (Chapter 2) extend the concept of produsage in citizen journalism, arguing that Twitter is a platform for news produsage in which many users make small and incremental contributions to the whole. As they put it, Twitter—especially via its hashtag and retweeting systems—"turbo-charges" gatewatching[5] practices by allowing users to share new or newsworthy information with the relevant user groups. In their quantitative analysis of selected hashtag streams, Bruns and Highfield demonstrate that the nature of citizen gatewatching changes, based on the availability of mainstream media coverage. Their investigation of manual retweets and tweets introducing new information revealed markedly different patterns between crisis or breaking news events and non-breaking or scheduled news events. The former category included a higher percentage of retweets and tweets including new information, reflecting extensive information gathering and sharing. Events in the latter category, on the other hand, were anticipated and covered by the mainstream media. Their analysis reveals an even higher presence of tweets including new information in the WikiLeaks hashtag stream. Bruns and Highfield note that Facebook, Twitter, and other such tools bring us ever closer to what Lasica called "random acts of journalism" (as cited in Bruns and Highfield, this volume). Besides further empowering "the people formerly known as the audience" (Rosen, 2006), these tools firmly underscore that journalism is solely the province of neither the professional nor the amateur.

Matt Hills (Chapter 7) focuses specifically on spoilers, or the release of information about a narrative (e.g., film, television program, game) prior to the narrative's official availability. His contribution presents an application of psychoanalysis to fans' self narrative that repudiates the discourses of pathology that have been, he argues, too often inappropriately linked to fan behaviors. Hills argues that spoilers pose emotional questions of anxiety, trust, and control. Besides their communal value, spoilers can work to contain ontological insecurity and help process it back into a sense of security. Avoiding the trap of pathologizing the emotions, anxieties, and conflicts of fans, and going beyond concepts of the unconscious and the technological unconscious, Hills presents fans as possibly enacting a "technological-narrative unconscious" in which their self-identity and their self-narrative as fans are protected.

Hills' contestation that certain behaviors have inappropriately been pathologized points to another theme evident across several readings. Taken broadly, this theme reflects a consideration of the psychological aspects of actors in our intersectional contexts, sometimes focusing on what is or is not pathological, and sometimes on the creation or the alleviation of anxieties. For example, Thiel-Stern's conceptualization of identity emphasizes that actors need not have only one fixed identity, an echo of Dreier's (2008) refutation of the pathologizing of the non-fully-integrated or non-fully-coherent personality—he argued, in fact, that for most people varied social practices are adaptive. Costa reminds us of the psychological defenses evident in Freud's binding and unbinding concepts. Yep, Olzman, and Conkle discovered that the videos they analyzed reveal the repression of what are called "ugly feelings," or affective responses which are important to acknowledge but are too frequently downplayed.

In addition, Catherine McGeehin Heilferty (Chapter 10) considers the psychological benefits of blogging about cancer. Her chapter presents one of two new theories being developed in this volume. Heilferty offers an early examination of her theory of online communication during illness. Building on her prior work, she conducts a thematic narrative analysis of blogs created by parents of children with cancer. She investigates the themes evident in the parents' narratives of illness, and the influence of author-reader interactivity expressed in the blogs. The themes in the blogs centered on what Heilferty calls "balance and ballast," with each theme invoking matched pairs of antithetical components: uncertainty and uncertainty management, stress and stress management, change and constants, burdens and gifts, and private lives and public lives. She found that the blogs ultimately were works of "cocreation," in that the blog authors expressed the need to hear regularly from the readers. Heilferty situates her results in a context of improved health care, participatory medicine, and nursing informatics, and acknowledges the need to understand more about the use and effect of not only social media but also the family as a whole when creating a plan of care.

A sense of belonging to a community, and a sense of identity,[6] are other important psychological components considered, not just in Heilferty's chapter and in the contribution by Lee and Webb, but throughout this volume.

Brittney Lee and Lynne Webb (Chapter 11) also report on a new theory under development. They surveyed mommy bloggers, to test the efficacy of their theoretical model depicting the mommy blogosphere in the United States. The Identity, Content, Community (ICC) model of blog participation predicts a total of eight relationships among six variables: identity, motivation, content, community, support, and relationships. Besides identifying two discrete groups of mommy bloggers, each blogging about a diverse but distinct range of topics, Lee and Webb's survey supports part of the ICC model by linking identity to blogging content and the bloggers' sense of community, and documents a link between blogging interaction—that is, writing and commenting on blogs, rather than merely reading blogs—and a sense of community in the blogosphere.

Taken together, these chapters illuminate many facets of the lives we, *Homo Irretitus*, lead. Focusing multiple spotlights onto the intersection of audiences and production made possible by social software helps make clearer a more nuanced perspective than would otherwise be possible. Yet it also reveals more questions—questions arising from each one of the various themes considered in this introduction as well as others. I hope you enjoy the process of finding answers, and uncovering questions, as you read what follows.

Notes

1. Noting the challenges of defining this concept, Coates qualified his definition as "loose" and as a "mostly accurate short-hand description of social software that mostly worked," but also said he might prefer it over an earlier, longer definition. The earlier definition (Coates, 2003) also referenced social software as "prosthesis."

2. Neil Postman's (1985) classic, *Amusing Ourselves to Death*, offers a fascinating discussion based on the future predicted by George Orwell in *1984* and that predicted by Aldous Huxley in *Brave New World*. Postman's excellently crafted foreword concludes by saying that his "book is about the possibility that Huxley, not Orwell, was right" (p. xix). So far, the intervening years have done little, if anything, to diminish that possibility.

3. For more on standpoint theory, see, for example, Harding (2004), Hartsock (1988) or Collins (1986).

4. Lesbian, Gay, Bisexual, Transgendered, Queer

5. For more on gatewatching, see Bruns (2005).

6. Unfortunately, space precludes fuller considerations of community and identity; for pivotal contributions see the work of Henry Jenkins (e.g., Jenkins, 2006a; 2006b), Nancy Baym (e.g., 2000; 2006; 2011), and Howard Rheingold (2000).

References

Barthes, R. (1977). *Image—Music—Text*. (S. Heath, Trans.). New York, NY: Hill and Wang. (Original work published 1967).

Baym, N. K. (2000). *Tune in, log on: Soaps, fandom, and online community*. Thousand Oaks, CA: Sage.

Baym, N. K. (2006). The emergence of on-line community. In S. Jones (Ed.), *CyberSociety 2.0: Revisiting computer-mediated communication and community* (pp. 35–68). Thousand Oaks, CA: Sage.

Baym, N. K. (2010). *Personal connections in the digital age*. Cambridge, England: Polity Press.

Berger, P. L. & Luckmann, T. (1967). *The social construction of reality: A treatise in the sociology of knowledge*. New York, NY: Anchor Press.

Bird, S. E. (2011). Are we all produsers now? Convergence and media audience practices. *Cultural Studies, 25*(4–5), 502–516.

Bourdieu, P. (1984). *Distinction: A social critique of the judgement of taste.* (R. Nice, Trans.) Cambridge, MA: Harvard University Press. (Original work published 1979).

Bruns, A. (2005). *Gatewatching: Collaborative online news production.* New York, NY: Peter Lang.

Bruns, A. (2008). *Blogs, Wikipedia, Second Life, and beyond: From production to produsage.* New York, NY: Peter Lang.

Coates, T. (2003, May 8). *My working definition of social software*...[Blog post]. Retrieved March 21, 2012, from http://www.plasticbag.org/archives/2003/05/my_working_definition_of_social_software/

Coates, T. (2005, January 5). *An addendum to a definition of social software* [Blog post]. Retrieved March 21, 2012, from http://www.plasticbag.org/archives/2005/01/an_addendum_to_a_definition_of_social_software/

Collins, P. H. (1986). Learning from the outsider within: The sociological significance of black feminist thought. *Social Problems, 33*(6), 14–32.

Dreier, O. (2008). *Psychotherapy in everyday life.* New York, NY: Cambridge University Press.

Fiske, J. (1989). Moments of television: Neither the text nor the audience. In E. Seiter, H. Borchers, G. Kreutzner, & E. Warth (Eds.), *Remote control: Television, audiences, and cultural power* (pp. 56–78). New York, NY: Routledge.

Giddens, A. (1984). *The constitution of society: Outline of the theory of structuration.* Berkeley, CA: University of California Press.

Giddens, A. (1991). *Modernity and self-identity: Self and society in the late modern age.* Stanford, CA: Stanford University Press.

Harding, S. (2004). Introduction: Standpoint theory as a site of political, philosophic and scientific debate. In S. Harding (Ed.), *The feminist standpoint theory reader: Intellectual and political controversies* (pp. 1–20). New York, NY: Routledge.

Hartsock, N. C. M. (1998). *The feminist standpoint revisited, and other essays.* Boulder, CO: Westview Press.

Heath, S. (1977). Notes on suture. *Screen, 18*(4), 48–76.

Jenkins, H. (2006a). *Convergence culture: Where old and new media collide.* New York, NY: New York University Press.

Jenkins, H. (2006b). *Fans, bloggers, and gamers: Exploring participatory culture.* New York, NY: New York University Press.

Lind, R. A. (2013). Laying a foundation for studying race, gender, class, and the media. In R. A. Lind (Ed.), *Race/gender/class/media 3.0:*

Considering diversity across audiences, content, and producers (3rd ed.) (pp. 1–12). Boston, MA: Pearson.

MacCabe, C. (1974). Realism and the cinema: Notes on some Brechtian theses. *Screen, 15*(2), 7–27.

Postman, N. (1985). *Amusing ourselves to death: Public discourse in the age of show business*. New York, NY: Penguin Books.

Rheingold, H. (2000). *The virtual community: Homesteading on the electronic frontier*. London, England: MIT Press.

Rosen, J. (2006, June 27). *The people formerly known as the audience* [Blog post]. Retrieved March 21, 2012, from http://archive.pressthink.org/2006/06/27/ppl_frmr.html

Saulauskas, M. P. (2000). The spell of *Homo Irretitus*: Amidst superstitions and dreams. *Information Research, 5*(4). Retrieved March 16, 2012, from http://informationr.net/ir/5-4/paper80.html

Unger, R. M. (2001). *False necessity: Anti-necessitarian social theory in the service of radical democracy* (new ed.). London, England: Verso.

CHAPTER TWO

Blogs, Twitter, and Breaking News: The Produsage of Citizen Journalism

Axel Bruns and Tim Highfield

Debates over the role and relevance of what has been described as citizen journalism have existed since at least the late 1990s; positions have ranged from the fulsome dismissal of such bottom-up journalism activities (and indeed, almost all user-led content creation) as being part of a new "cult of the amateur" (Keen, 2007) to nearly equally simplistic perspectives that predicted citizen journalists would replace the mainstream journalism industry within a short timeframe. A more considered, more realistic perspective would take a somewhat more moderate view. Aided by circumstances including the long-term financial crisis enveloping journalism industries in many developed nations, the creeping corporatization and politicization of journalistic activities in democratic and non-democratic countries alike, and the largely unmet challenge of new, Internet-based media forms, citizen journalism (as well as other parajournalistic media, including TV comedy such as *The Daily Show*) has been able to make credible inroads into what used to be the domain of journalism proper.

Technology has played an important role as disruptor and enabler in these developments, even if—of course—it has not determined their eventual course. First, the rise of the Internet as a popular medium has led to a substantial increase in available channels for information and entertainment, among other purposes. One such further purpose, indeed, has fundamentally undermined the existing business model of conventional newspaper publishing; with specialist Web sites and generalist search engines providing a more effective and easily searchable platform for job, real estate, car and other advertisements, and with the subsequent shift of such advertising to new and independent online platforms, much of the financial foundation of newspaper journalism has been eroded beyond recall. Coupled with broader economic trends, this shift alone has led to the decline of many once well-established newspapers.

Second, and relatedly, the proliferation of possible channels for news content, and the increasing precariousness of their financial bases, have combined to make corporate pressures on fearless, independent journalism felt ever more immediately. On the one hand, this simply means that few journalistic organizations can afford to engage in much long-form, resource-intensive, investigative journalism, and that press release journalism that barely conceals news stories' origins in commercial and government releases has grown correspondingly. Such tendencies lower the average quality of journalistic publications and broadcasts, and thereby further undermine the attractiveness of the journalistic product, leading to a continuation of the de-cline in audiences. On the other hand, the flipside of A. J. Liebling's famous statement "freedom of the press is guaranteed only to those who own one" (1960, p. 105) is revealed: potential commercial returns now only rarely still constitute the principal motivation for owning and operating a newspaper or news channel—rather, with owning a press organization comes the opportu-nity to freely engage in political lobbying (or indeed, pressure) in favor of the proprietor's political or corporate causes. Thus, influential newspapers such as *The Australian* or *The Times* are unlikely ever again to be able to fund their own operations from sales and advertising; rather, they are cross-subsidized by the other, more lucrative branches of the conglomerate that owns them, and they repay that investment in the currency of political influ-ence. This, however, also means that editorial commentary (at least), as well perhaps as journalistic content (often), will and must be slanted toward these selected causes—freedom of the press might be guaranteed, but freedom of the individual journalist isn't.

Third, partly because of the already depleted resources available to mainstream journalistic operations, partly because of a deep-set conservatism among journalistic staff, and partly also because many possible new models of operation are inherently anathema to the long-established journalistic way of doing things, the journalism industry has been slow to embrace the poten-tial offered by new, Internet-based media forms. Many early online news of-ferings by established news organizations limply replicated part of the content available in their corresponding newspapers (in formats ridiculed as "shovelware"); even well into the 2000s and what is often (although not un-problematically) called the "Web 2.0" era, few mainstream news Web sites sought to position their readers as anything other than largely passive audiences, or to directly engage and correspond with them. Readers as potential sources of story ideas or feedback were often addressed only in a tokenistic, haphazard fashion; reader comments (attached to news stories or op-eds) became more commonplace, but often still provided little more than a space for random rants which were neither policed nor engaged with by journalistic or editorial staff. It is only recently, and largely as a result of the growing competitive pressure from alternative, citizen journalism Web sites, that more advanced approaches to interaction are being tested in earnest, and even then only by a handful of news organizations—famously, for example, *The Guardian* (2011) sought its readers' collaboration in sifting through the tens of thousands of documents related to the UK MPs' expenses scandal, providing an interface for readers to annotate and highlight information in any of these documents and thereby help uncover further instances of corruption and wrongdoing.

Citizen Journalism?

These shortcomings of industrial journalism combined with the opportunities inherent in new Internet-based media forms and platforms to give rise to the new models that have been described as "citizen journalism." To some extent, the term is a misnomer: it implies that professional journalists are not *also* citizens (meaning in this context, invested in the future political and societal course of their country), and it equates the news-related activities of "citizen journalists" with the journalism undertaken by professional staff in the news industry—strictly speaking, both these assumptions are incorrect.

While the former point may be self-evident, the latter requires some further discussion. Overall, "citizen journalism" refers to an assemblage of broadly journalistic activities (what J. D. Lasica [2003, p. 73] has described as "random acts of journalism") that are characterized by specific practical and technological affordances: they draw on the voluntary contributions of a wide-ranging and distributed network of self-selected participants rather than on the paid work of a core team of professional staff, and they utilize Internet technologies to coordinate the process and share its results. Such activities may take place under the auspices of a central Web site (from *Indymedia* through *Slashdot* and *OhmyNews* to the *Huffington Post*), or unfold in a more decentralized fashion through interactions among individual participants in the (political) blogosphere or through the collaboration of dispersed networks of individuals using a shared underlying social media platform such as Twitter.

Arguably, citizen journalism began as a direct and determined response to the perceived shortcomings of mainstream journalistic coverage; often cited as a seminal moment, for example, was when the first *Indymedia* site was established to cover the 1999 World Trade Organization meeting in Seattle and the political protests surrounding it, which became known as the "Battle of Seattle" (see Meikle, 2002). Activists, who anticipated that mainstream media coverage of their protests would be strongly biased toward portraying them as anarchists and hooligans, took matters into their own hands by publishing their own, alternative text, audio, and video reports from the protests through the then still new digital publishing platform of the Web. Similar alternative reporting from such events has since become a standard feature of political protests, and has effectively undermined the ability of politicians and mainstream media alone to determine how such protests are framed in news reports.

This orchestrated firsthand reporting from major events remained somewhat unusual over the next few years, however—not least because of the planning and organization required. Rather, much of the citizen journalism that rose to prominence in the early 2000s could be better described as "citizen commentary": independent responses to political events and developments that provided an alternative, bottom-up view to the top-down and sometimes self-censoring narrative of mainstream journalism. Such citizen commentary was especially important in the aftermath of the events of September 11, 2001, when few U.S.-based (and even international) news media organizations dared to openly criticize the belligerent response of the Bush administration or the conduct of the subsequent wars in Iraq and Afghanistan, for fear of being branded "unpatriotic." Citizen commentary—as well as a handful of new news-related formats, especially including news satire television such as *The Daily Show*—filled the gap created by self-censorship in the news industry, and such alternative news media (many of them blogs or blog-style collaborative sites) became established elements in the wider news and political media ecology.

As spaces for news *commentary*, such sites necessarily operate somewhat differently from sites built more strongly around firsthand alternative reporting, such as *Indymedia* or *OhmyNews*, even if the latter also carry commentary and share some basic traits. Commentary responds to existing, already published news and opinion; it collects, collates, and combines these existing materials, contextualizes them and thereby points out new frames for their interpretation and analysis. Commentary, in short, compiles information and can therefore be seen as a form of news curation: of tracking a story and highlighting its origins and implications.

Additionally, of course, commentary expresses an opinion, and therefore violates the journalistic ideal of objectivity (a fact that has resulted in bloggers and other independent commentators being described at times as "armchair journalists" or worse). However, it should be pointed out in this context that objectivity remains little more than an ideal in professional journalism, too, and has been deeply compromised in practice by political and corporate pressures. Indeed, objectivity (or at least its more achievable cousin, balance) is an aim that was more important to journalistic activities at a time of comparable channel scarcity; in a mediasphere characterized by an abundance of channels, it has been replaced by the ideal of multiperspectivality, as Herbert Gans outlined it:

> Ideally, ...the news should be omniperspectival; it should present and represent all perspectives in and on America. This idea, however, is unachievable, for it is only another way of saying that all questions are right. It is possible to suggest, however, that the news, and the news media, be multiperspectival, presenting and representing as many perspectives as possible—and at the very least, more than today. (1980, pp. 312–313)

Taken as a whole (rather than individually for each Web site), then, citizen commentary in combination with the news coverage of the mainstream media provides for such multiperspectivality.

News curation is the core practice of these bottom-up, independent, participatory citizen journalism and citizen commentary sites. There can no longer be any suggestion that such sites act as traditional journalistic gatekeepers; they inherently and necessarily draw on materials already published by other mainstream and alternative news and opinion sites, to curate and comment on them. The same, however, is also true for the mainstream news media now: following the multiplication of channels that resulted especially from the emergence of the World Wide Web as a popular medium, their publications (online as well as offline), too, no longer perform a significant gatekeeping role. Excepting the work of deep investigative journalism (which, sadly, is a declining art), most of the information that even industrial journalism draws on is already available to the public, through press releases and statements from government, commercial, and non-government organizations. The gates have multiplied beyond all control, and nobody is able to keep them any more.

What replaces gatekeeping, then, can usefully be described as *gatewatching* (Bruns, 2005): professional and citizen journalists and commentators watch the gates of newsworthy organizations whose information is relevant to their specific interests; they capture and compile that information as it is released; and they process and curate such information with an aim to publish news stories and comments that build on it. Professional journalists may still be able to follow up by calling sources and spokespeople for additional statements, while such courses of action may be unavailable to unresourced non-professionals; professional gatewatching therefore results in a comparatively greater number of news stories, whereas bottom-up, independent gatewatching tends to focus more strongly on compiling and commenting on available information. Gatewatching itself, however, is a core practice for both groups.

In principle, what may result from this could be a relatively straightforward split between professional news journalism, published by the news industry, and alternative news commentary, expressed through blogs and other independent Web sites and platforms. However, the politicization of mainstream journalism and the relatively low cost of producing commentary (compared to news reports) have also led to substantial growth of the space reserved for news commentary in mainstream news outlets in print, broadcast, and online (and this, in turn, fuels alternative news sites' desire to provide independent information that responds to what they perceive as political spin in the mainstream news). As a result, practices in mainstream and alternative news and commentary now overlap considerably, with such interconnections also evident in the sharing of personnel. In Australia, for example, all three major news organizations have launched dedicated online spaces for news and political commentary: the Murdoch-owned News Corp set up *The Punch*, the Fairfax Media group launched *The National Times*, and public broadcaster ABC relaunched an existing commentary space as *The Drum*— and in each case, the contributors include seasoned political journalists as well as leading Australian news and politics bloggers.

As news and commentary blend, and as professional journalists and independent commentators can no longer avoid engaging with one another, the potential exists for a new public sphere to emerge in and through these spaces. Notably, this does not imply that such a public sphere will necessarily be *functional*; the senselessly reductionist "us vs. them" arguments that have characterized many of the early debates about citizen journalism (see, for example, Bruns, 2005; Rosenberg, 2002) may well continue. At the very least, however, the underlying structure of the situation has changed considerably since the early 2000s. Previously, industrial journalists and citizen journalists may have remained at some distance from one another, with few direct personal contacts; now they well and truly occupy the same shared space, as a heterogeneous group if not a homogeneous community.

A Shared Space of News Produsage

This shared space now extends across a variety of online platforms (as well as into the offline world). It contains the Web sites of mainstream news organizations as well as the news and politics blogs and citizen journalism sites of independent citizen journalists and commentators; additionally, it takes in the social networks of Facebook, Twitter, and elsewhere its protagonists—from professional journalists through independent commentators to those committing the rare "random act of journalism"—can be found.

Indeed, the emergence of these latter social networking spaces during the second half of the 2000s has yet again served to expand the range of participants available to engage in news dissemination, curation, and commentary. Previously, citizen journalism (or at least, citizen commentary) still meant setting up one's own blog or blog-style site, or becoming a regular contributor to an existing site; now, little more than a user registration on Facebook or Twitter is required to become engaged in news and political commentary and even to rise to some degree of prominence for doing so. The short messaging service Twitter, in particular (more than Facebook), provides an unprecedentedly open and accessible space for such activities. It it built on a much simpler networking structure where updates posted by users are either public or private, rather than visible and shareable only to selected circles of friends within one's social network; indeed, public Twitter messages are visible even to unregistered visitors to the Twitter Web site.

Additionally, and building on this flat and open network structure, Twitter users have instituted the hashtag system (brief keywords preceded by the symbol, "#"), which provides a simple and elegant solution for tagging one's own updates as relevant to specific topics, and for thereby making such updates (and the continuing discussions to which they may belong) discoverable and traceable by others. Hashtags enable public conversations by large groups of Twitter users without each participating user needing to subscribe to (to "follow") the update feeds of all other participants; they are also especially effective at establishing topical communities ad hoc, such as in response to breaking news stories. Hashtags are used for a wide variety of purposes, in fact: from breaking news (such as #tsunami, for the March 2011 earthquake and tsunami in Japan) through continuous discussion and regular events (such as #auspol for political discussion in Australia, or #qt for discussion of events in Prime Minister's Question Time in the Australian parliament), to more phatic and emotive uses that are unlikely to be intended as a means of coordinating ongoing conversations (such as #tired, #fail, or

#headdesk). For the purposes of our present discussion, however, the use of hashtags to mark contributions to a topical discussion and to coordinate processes of exchanging information and opinion are of particular interest.

Twitter, then (as well as other social media platforms that may be used in a similar fashion), further extends the potential participant base for citizen commentary (and, to some extent, also for citizen journalism proper). In the first place, it in effect turbo-charges the practices of gatewatching: as soon as new and newsworthy information is discovered to be passing through the gates of a relevant organization, it (and/or hyperlinks to the source) can be shared with the wider Twitter user base, hashtagged to be especially visible to relevant existing user groups. Additionally, more- or less-considered discussion and commentary on the new information can ensue, and users can compare and connect the new information with other available material. In this way, the process of news curation that we have already encountered in previous citizen journalism or citizen commentary practices (but which in those contexts often remained the domain of the individual blogger or commenter) is further decentralized and shared; no one individual Twitter user is now responsible for compiling, collating, and curating the available information on any given topic. Instead, it becomes a thoroughly collaborative exercise.

As a result, Twitter comes yet closer to Lasica's (2003) idea of "random acts of journalism" than was the case for earlier alternative online media forms. Not least simply because of the 140 character limit, individual tweets by individual users must *necessarily* constitute fragments of journalistic activity that make full sense only in combination and in the context of the activities of other participating Twitter users. Describing Twitter as a platform and medium for "ambient journalism," Alfred Hermida (2010) and Alex Burns (2010) have both drawn parallels between Twitter's coverage of newsworthy events and the patterns of ambient music, in fact: much as musical form in ambient music only emerges over time from often minimal variations in tone, timbre, and rhythm, so too does journalistic coverage and commentary on Twitter only emerge from the sum of a larger number of individual tweets containing summary information, links to further materials, and user comments and evaluation.

In this, then, the processes of news coverage and discussion on Twitter, even more than the back-and-forth exchanges in the news and political blogosphere or between blogs and mainstream journalism, position Twitter as a platform for news produsage (Bruns, 2008a): that is, the gradual and collaborative development of news coverage and commentary by a wide range of users voluntarily making small and incremental productive contributions to the whole, rather than the orchestrated production of news stories and opinion by small teams of dedicated professionals. Twitter's underlying social, organizational, and technological structures make it exceptionally easy for users to participate in such ambient journalism processes: all that is required is that they post a short message containing the information or opinion they intend to share, as well as perhaps to mark it with a hashtag in order to increase its visibility to a wider target audience. In other words, participation in news dissemination, curation, and commentary processes on Twitter is open to all comers; through their random acts of journalism, Twitter participants are neither simply users nor fully producers of news coverage, but placed in a hybrid role as *produser*; and whether the contributions made by any individual user have any impact depends on the reaction and evaluation by other users, and especially on their sharing and further dissemination of such contributions through retweets (passing on these messages to their own followers either verbatim or in an edited, commented, contextualized form).

What gradually emerges from these processes of sharing and resharing on Twitter, in turn (as well as from overall patterns of following and being followed), are both more permanent and more ad hoc community structures: more permanent structures of influence as users follow others, where those users who have the most followers are most easily able to disseminate their information and opinion widely through the network; more ad hoc structures through interactions within topical hashtag communities, where for the duration that a hashtag is active specific users messages may be most widely visible as a result of the retweeting and responding activities of others. Such ad hoc patterns, especially, may shift more or less quickly over time, as new users with important new information join the discussion, or as others have no further ideas or comments to add to the current debate.

Somewhat paradoxically, the resultant knowledge (as well as interpretation and opinion) base which is established and continuously maintained through these processes exists everywhere and nowhere at the same time. Dedicated followers of the hashtag discussion can develop for themselves a comprehensive understanding of the issues and views involved; others who dip in and out of the discussion or track it only remotely as followers of selected contributors to the hashtag community will only see a partial representation of the full range of exchanges. Twitter remains, in the first place, a discovery mechanism for news and views, rather than any kind of full replacement for more comprehensive platforms for news coverage and commentary—and the news and views that are retweeted and discussed most frequently will reach the largest number of Twitter users. At the same time, of course, it would be a mistake to regard Twitter in isolation from other such platforms. It is itself both a disseminator of links to further information elsewhere (especially in other online media), and embedded in a growing range of other media forms and formats—from Twitter feeds on selected topics or by selected authors as they appear on the pages of blogs and online newspapers, to superimposed Twitter messages incorporated into the content of mainstream news channels. In many ways, Twitter provides the glue—the interconnections—between a wide and diverse variety of different news sources and platforms, and these interconnections are driven by Twitter users' gatewatching efforts.

From Ambient to Actual: Breaking News

Such gatewatching efforts are especially visible during breaking news events. Breaking news has always posed a challenge for news organizations, and the pressures of the 24-hour news cycle have further amplified these challenges; real-time online and social media such as Twitter add new complications as information purporting to originate from affected local areas is often immediately available, but has yet to be independently verified. Twitter's (and more broadly, social media's) utility for the live coverage of breaking news—and the journalistic problems that come with such coverage—have been well-identified since at least the first major breaking news story closely identified with Twitter, the emergency landing of a commercial flight on New York's Hudson River in 2009. The incessant string of natural disasters and political upheavals in 2011, the coverage of each of which Twitter played a significant role in, has further cemented these perceptions (see, for example, Lotan et al., 2011, for a first examination of the use of Twitter in the Arab Spring uprisings).

Breaking news events demonstrate especially clearly the importance of Twitter as a mechanism for distributed gatewatching. As Twitter users encounter early rumors about an unfolding newsworthy event (often via Twitter itself), some will search for further information, and share their findings with the wider Twitter community in turn. Some will also include what they deem an appropriate hashtag for the event. This, then, leads to further users encountering such information (and such hashtags), and a proportion of these users will again become active in seeking and sharing additional information, as well as in retweeting the material already shared by their predecessors; as a result, coverage of the event, and its corresponding hashtag(s), gain prominence on Twitter as a whole as well as in the update feeds received by individual Twitter participants from the users they follow. (Twitter provides a simple indicator of what it calls "trending topics," both global and organized by specific geographic regions: a list of hashtags and keywords whose over-all volume of mentions has increased especially rapidly during the preceding hours.)

In addition to the sharing and retweeting of relevant information about the breaking news event, commentary and discussion that evaluates the available information and begins to shed light on contexts and backgrounds to the story also emerges rapidly; in the process, the real-time communication activities taking place on the Twitter platform provide not so much a "first draft of history," as journalism has been famously described, but in essence a first draft of the present, to be revised and completed as further information comes to hand. Distinctions between rumor and fact are especially crucial at this stage, and are attempted through crowdsourced, collaborative processes, with varying degrees of success; similarly, and in direct connection to such verification processes, leading participants in the discussion and curation of information relating to the breaking news event are also gradually identified. In the context of natural disasters such as the 2011 southeast Queensland floods or the 2010 and 2011 earthquakes in Christchurch, New Zealand, the ecosystem of leading information sources on Twitter contained a range of emergency authorities, mainstream media organizations, NGOs, and directly affected local users (Bruns, Burgess, Crawford, & Shaw, 2011; 2012).

The effectiveness of such information dissemination and evaluation activities also depends on Twitter users avoiding the fragmentation of their efforts by settling on a single, widely used hashtag. As noted, when news breaks, participating users will usually experiment with a range of possible hashtag solutions; in early Twitter coverage of breaking news about the 2011 Oslo bombing by political extremist Anders Breivik, for example, hashtags ranging from #oslo through #osloexpl to #oslobomb could be observed before news about the massacre on Utøya island shifted the focus toward the more encompassing hashtag #norway. In contrast, during the second and third major earthquakes in Christchurch, in February and June 2011, Twitter users recycled the #eqnz hashtag which had already been used during the first event in September 2010 (Bruns & Burgess, 2011a). Hashtags can be used to facilitate the formation of ad hoc publics on Twitter, but whether such processes are successful depends on the extent of uptake by the larger community of participating users (Bruns & Burgess, 2011b).

Overall, however, the coverage of breaking news events on Twitter tends to demonstrate the point at which ambient journalism becomes actual news coverage and dissemination, where a substantial number of Twitter users come together to—in journalistic parlance—"work the story" and engage in a more or less committed and orchestrated effort at gatewatching and disseminating relevant information. Indeed, in line with the view that citizen journalism and user-led news produsage are driven at least in good part by the perceived shortcomings of mainstream journalism, such activities appear to be especially focused in breaking news contexts, where news organizations may still be scrambling to get their reporters to the scene and find footage of the events; as and when sufficient mainstream media coverage is available, gatewatching efforts by the Twitter community decline, and shift focus toward commentary and evaluation rather than news dissemination.

This is clearly evident from a quantitative analysis of tweet types within selected hashtag streams addressing foreseen and unforeseen newsworthy events, as depicted in Figure 2.1: here, we examine the percentage of tweets in each overall hashtag dataset that contain links to external sources—URLs—and/or unedited manual retweets (i.e., retweets without additional commentary by the retweeting user). Together, these two metrics indicate the extent to which participating users are both introducing new information to Twitter (URLs) and disseminating existing information verbatim to their own follower networks (retweets).

In our analysis, two obvious clusters of similarly structured hashtag conversations emerge: on the one hand, a cluster containing immediately crisis- and breaking news-related hashtags, ranging from the 2011 southeast Queensland floods (#qldfloods) through the 2010 and 2011 Christchurch earthquakes (#eqnz) and the 2011 Japanese tsunami (#tsunami) to political upheavals including the 2011 revolution in Libya (#libya) and the 2011 riots (and subsequent, self-organized cleanup efforts) in London and the UK (#londonriots, #ukriots, and #riotcleanup). Each of these hashtag datasets—collected from the early stages of the event through to its relative conclusion, when overall tweet volumes had dropped considerably—is marked by a substantial percentage of URLs in tweets (above 40% on average) and retweets (between 50% and 60% on average), indicating significant sourcing and sharing of information. This constitutes clear and direct evidence of gatewatching processes in action, responding to an urgent need for more information within minutes of first news or rumors of the acute event itself, as well as each time that further aspects of the event begin to unfold (or, especially in the case of #libya, throughout a lengthy unfolding chain of acute events).

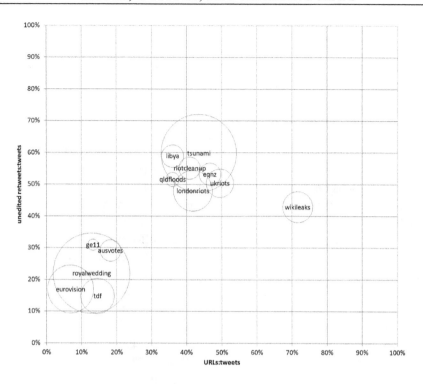

Figure 2.1: User activity patterns for selected Twitter hashtags (size indicates total number of contributors)

By contrast, a second cluster of hashtags is marked by a substantially lower percentage of URLs in their respective datasets (between 10% and 20% on average) and a similarly smaller percentage of retweets (roughly between 10% and 30% on average). The hashtags collected here represent a diverse group of events ranging from major cultural and sporting events such as the 2011 Eurovision Song Contest (#eurovision) and the 2011 Tour de France (#tdf) to the 2011 British royal wedding (#royalwedding) and the 2011 Australian and Irish general elections (#ausvotes and #ge11, respectively—with Twitter activity strongly skewed toward election day in each case). These hashtags address non-breaking news events: scheduled events whose dates had been known well in advance, and for which ample live television and other media coverage was available. Here, clearly, there is very little need for sustained gatewatching practices (even if some Twitter users may feel that specific features or aspects of these events are underreported in the mainstream media), manifesting in a much reduced incidence of information sourcing and sharing practices. Instead, the activities taking place here are much more closely related to the cultural studies concept of *audiencing* (see, for example, Fiske, 1992): participation in the shared experience of engaging with the same media text together (if at a distance), at the same time, as members of an imagined community of audience members. Finally, the #wikileaks hashtag, used for discussion of the continuing controversies surrounding the *WikiLeaks* whistleblower Web site and its founder, Julian Assange, acts as a further outlier to these two clusters of similarly structured hashtags. Its patterns of user activity appear somewhat similar to the hashtags collected in the breaking news cluster, but #wikileaks contains a substantially greater percentage of tweets with URLs (over 70%), indicating an even greater focus on identifying and sharing information available from external sources—this may be an indication of the fact that *WikiLeaks* supporters believe that their causes are especially poorly reported by the mainstream media, and that they are engaging in a concerted effort to spread relevant information through other means. Additionally, of course, the very purpose of *WikiLeaks* is to disseminate previously unavailable materials to the public; an especially high incidence of URLs is therefore hardly surprising in this case (for more on #wikileaks, see Lindgren & Lundström, 2011).

It is also striking that these patterns appear to remain constant regardless of the total number of unique Twitter users contributing to each hashtag (the size of the overall user community contributing to each hashtag is indicated in Figure 2.1 by the size of each node on the graph). Only some 9,500 users participated in the #ge11 discussion of the 2011 Irish general election during its final days, for example, but their patterns of tweeting are scarcely different from the 15,000 participants in the #ausvotes discussion of the Australian federal election during July and August 2010, or from the more than 502,000 users posting tweets containing the #royalwedding hashtag. Similarly, the 15,500 contributors to #qldfloods or the 38,500 posters to #eqnz share URLs and retweets at a rate comparable to the 474,000 Twitter users posting tweets containing the #tsunami hashtag. Audiencing or gatewatching practices, it seems, do not depend on the existence of a substantial critical mass of fellow participants, at least once a common hashtag has been established.

Conclusion: Blogs, Social Media, and the Produsage of Citizen Journalism

Digital environments have changed substantially since the emergence of the first online citizen journalism projects in the late 1990s. Painstakingly hand-coded platforms for the collaborative identification, publishing, and discussion of current issues and stories, such as *Indymedia* and *Slashdot*, have given way to more advanced Web 2.0 tools and social media spaces, but the underlying principles of such activities appear to have remained remarkably constant. Gatewatching continues to be the core practice supporting and enabling the produsage of quasi-journalistic content by users for users; whether on blogs, in collaborative news platforms, or through collective action in social networking services, interested users are engaging in collective processes of finding and disseminating newsworthy materials from a wide variety of sources, and in collaboratively curating these collections and interpreting their content.

The overview of crisis-related gatewatching activities on Twitter presented here can only provide a few general insights into these practices. Further, more detailed studies are needed to explore the nature of the source materials being shared on Twitter and through other spaces (Highfield, 2011), for example, the community dynamics of the loose agglomerations of participating users which form and dissolve again around shared topics of interest (Bruns, 2010a; 2010b), the temporal processes of the coverage of acute news events on social media platforms, or the breadth of possible news events that operate according to the patterns we have observed here. There appear to be clear differences in Twitter's reaction to breaking, unforeseen news and to foreknown, widely televised events, for example, but to what extent is a theme such as *WikiLeaks* an outlier from these two major groups of events, or an example of a third major category of stories: long-term controversies that are to some extent ignored by the mainstream media?

But well beyond Twitter as a current space for what we continue to call, somewhat loosely, "citizen journalism," the practices of user-led sharing, reporting, and discussing of news stories appear to be alive and well; indeed, social media spaces such as Twitter indicate that participation in these practices—that is, committing "random acts of journalism"—has become even easier in recent years. In general, online and social media that allow for substantial user participation and collaboration inherently carry an opportunity to move from ambient news sharing to actual citizen journalism, as specific stories happen to, or near, users who are able to cover them firsthand, or as such stories attract the attention of a number of participants sizable enough to make existing reports more widely visible through sharing and commenting activities.

What is slowly changing, however, is the relationship between citizen and professional journalists. Clear structural boundaries between the Web sites of news organizations and the blogs and collaborative spaces of citizen journalism enabled the maintenance of an "us vs. them" attitude that manifested in a series of often highly acrimonious "blog wars" between the journalism industry and its independent, upstart critics (see, for example, Bruns, 2008b) in the first years of the new millennium; clichés of bloggers as pajama-clad armchair critics of hardworking professional journalists emerged during this time. In contrast, Twitter and other social media platforms constitute a more neutral space, shared by professionals and amateurs (or more accurately, peopled by users with a wide range of professional backgrounds and diverse forms of personal expertise), making it much more difficult for defenders of the true journalistic faith to attack and dismiss that space out right. Here, many (though not all) participating journalists engage more freely with their supporters as well as their critics, and—especially in the context of breaking, acute news events—do not shy away from drawing on these other users as potential sources for their stories. The blog wars have given way to a wary truce, at least, if not to an outright peace.

Such dissolution of the battle lines cannot help but change the journalistic industry as well as its position in society. Journalists' embrace of social media to disseminate, discuss, and expand their coverage of specific stories turns those stories from finished products into the unfinished, evolving artifacts common to produsage processes, and invites broader citizen engagement and participation in them, even if the Web sites of their news organizations still claim to contain the finished product, the complete article. Users who access (and perhaps share) these articles, as well as discuss, compare, and curate the overall coverage of an event or story through their participation in various online and social media platforms, sometimes in direct communication and exchange with the journalists involved, already know better; they know to look to their social networks (understood here in the narrow technical sense as well as the wider societal sense) for context, confirmation, or contestation. The journalism—the collaborative pro-am journalistic coverage—that emerges from this is a shared journalism, one which no longer belongs to news organizations or news audiences alone. As Herbert Gans once suggested, "the news may be too important to leave to the journalists alone" (1980, p. 22)—and in social media environments where news is ambient, shared, fluid, and circulating, it no longer is.

References

Bruns, A. (2005). *Gatewatching: Collaborative online news production.* New York, NY: Peter Lang.

Bruns, A. (2008a). The active audience: Transforming journalism from gatekeeping to gatewatching. In C. Paterson & D. Domingo (Eds.),

Making online news: The ethnography of new media production (pp. 171–184). New York, NY: Peter Lang.

Bruns, A. (2008b). *Blogs, Wikipedia, Second Life and beyond*: From production to produsage. New York, NY: Peter Lang.

Bruns, A. (2010a, December 30). Visualising Twitter dynamics in Gephi, part 1 [Blog post]. Retrieved August 1, 2011, from http://mappingonlinepublics.net/2010/12/30/visualising-twitter-dynamics -in-gephi-part-1

Bruns, A. (2010b, December 30). Visualising Twitter dynamics in Gephi, part 2 [Blog post]. Retrieved August 1, 2011, from http://mappingonlinepublics.net/2010/12/30/visualising-twitter-dynamics -in-gephi-part-2

Bruns, A., & Burgess, J. (2011a, October). *Local and global responses to disaster: #eqnz and the Christchurch earthquake.* Paper presented at the Association of Internet Researchers Conference, Seattle, WA.

Bruns, A., & Burgess, J. (2011b, August). *The use of Twitter hashtags in the formation of ad hoc publics.* Paper presented at the European Consortium for Political Communication Conference, Reykjavík, Iceland.

Bruns, A., Burgess, J., Crawford, K., & Shaw, F. (2011, April). *Social media use in the Queensland floods.* Paper presented at the Eidos Institute symposium Social Media in Times of Crisis, Brisbane, Queensland.

Bruns, A., Burgess, J., Crawford, K., & Shaw, F. (2012, January). *#qldfloods and @QPSMedia: Crisis communication on Twitter in the 2011 South East Queensland floods.* Brisbane, Queensland: ARC Centre of Excellence for Creative Industries and Innovation. Retrieved February 8, 2012, from http://cci.edu.au/floodsreport.pdf

Burns, A. (2010, May). Oblique strategies for ambient journalism. *M/C Journal, 13*(2). Retrieved August 1, 2011, from http://journal.media-culture.org.au/index.php/mcjournal/article/view/230

Fiske, J. (1992). Audiencing: A cultural studies approach to watching television. *Poetics, 21*(4), 345–359.

Gans, H. J. (1980). *Deciding what's news: A study of CBS Evening News, NBC Nightly News, Newsweek, and Time.* New York, NY: Vintage.

Hermida, A. (2010, May). From TV to Twitter: How ambient news became ambient journalism. *M/C Journal, 13*(2). Retrieved August 1, 2011, from http://journal.media-culture.org.au/index.php/mcjournal/article/view/220

Highfield, T. (2011). *Mapping intermedia news flows: Topical discussions in the Australian and French political blogospheres* (Unpublished doctoral thesis). Queensland University of Technology, Brisbane, Queensland.

Investigate your MP's expenses. (2011). *The Guardian.* Retrieved September 21, 2011, from http://mps-expenses.guardian.co.uk/

Keen, A. (2007). *The cult of the amateur: How today's Internet is killing our culture and assaulting our economy.* London, England: Currency.

Lasica, J. D. (2003). Blogs and journalism need each other. *Nieman Reports,* 70–74. Retrieved March 5, 2012, from http://www.nieman.harvard.edu/reports/03-3NRfall/V57N3.pdf

Liebling, A. J. (1960, May 14). Do you belong in journalism? *The New Yorker,* 105.

Lindgren, S., & Lundström, R. (2011). Pirate culture and hacktivist mobilization: The cultural and social protocols of #WikiLeaks on Twitter. *New Media & Society, 13*(6), 999–1018.

Lotan, G., Graeff, E., Ananny, M., Gaffney, D., Pearce, I., & boyd, d. (2011). The Arab Spring: The revolutions were tweeted: Information flows during the 2011 Tunisian and Egyptian revolutions. *International Journal of Communication, 5,* 1375–1405. Retrieved September 28, 2011, from http://ijoc.org/ojs/index.php/ijoc/article/view/1246/613

Meikle, G. (2002). Future active: Media activism and the Internet. New York, NY: Routledge.

Rosenberg, S. (2002, May 10). Much ado about blogging. *Salon.* Retrieved September 27, 2004, from http://www.salon.com/tech/col/rose/2002/05/10/blogs/

CHAPTER THREE

Procedure and Performance in an Era of Digital Media

Jay David Bolter

The defining moment in Christopher Nolan's film *The Dark Knight* is not a scene filled with violence and computerized special effects. It is a relatively quiet conversation in a hospital room between the Joker and Henry Dent, whose girlfriend the Joker has killed through a series of machinations. When Dent accuses him of having planned the woman's death, the Joker replies:

> Do I really look like a guy with a plan? ...You know, I just...do things. The mob has plans, the cops have plans, Gordon's got plans. You know, they're schemers. Schemers trying to control their little worlds. I'm not a schemer. I try to show the schemers how pathetic their attempts to control things really are. ...You were a schemer, you had plans, and look where that got you. I just did what I do best. I took your little plan and I turned it on itself. ...You know what I've noticed? Nobody panics when things go according to plan. Even if the plan is horrifying!...Introduce a little anarchy. Upset the established order, and everything becomes chaos. (Nolan, 2008)

The Dark Knight is an eclectic film, popular and serious at the same time. It presents itself as a conventional action-adventure movie and yet manages to incorporate themes associated with the avant-garde in the 20th century. Nolan's Joker is both a comic-book figure and a sort of Fluxus artist who stages violent happenings. He is a master planner and yet claims to have no plan. Throughout the film, the Joker configures objects and unwitting human victims into elaborate mechanisms, rather like Rube Goldberg machines, whose components fall like dominos at the appropriate moment and lead to more than a little anarchy.

Schemers and schemes are not limited to an imaginary Gotham City. Nolan was invoking and critiquing a notable feature of contemporary media culture: procedurality. Like the Joker, we as a culture seem to be fascinated by the ways in which our lives are governed by procedures. While imposing procedures on the populations of industrialized countries is by no means new, we are now introducing the peculiar character of digital technologies into more and more aspects of our daily life, by collective decision and individual preference. Individuals and groups sometimes recoil at this process, and the reaction against proceduralization may win temporary, regional victories. For example, the EU countries may adapt laws aimed at protecting their citizens from Internet giants such as Google collecting data in order to deliver them as commodities to other companies for targeted advertising. In the United States, politicians question the use of database and data mining by law enforcement and in anti-terrorism efforts. (Typically, Europe feels that the danger of abuse through digital procedurality comes from private companies, whereas the United States shows little concern about the private sector and instead frets over government abuse.) In any case, these limited concerns do not seem to impede the ingenuity of digital culture in finding new departments of our mediated daily lives to proceduralize.

The Mechanical and the Procedural

Procedurality is the latest phase in the mechanization of advanced industrial society; the process of social and cultural accommodation to industrial mechanization has extended over hundreds of years, as historians of technologies, such as Lewis Mumford in *Technics and Civilization*, have long argued. Mumford's (1934/2010) ambitious history, for example, divided the process into three periods, from the 15th century to the 20th: the eotechnic (represented or symbolized by the clock), the paleotechnic (iron and steam), and the neotechnic (electricity). Although a grand descriptive theory such as Mumford's is too simple to capture the nuances, it is clear that the history of mechanization in Europe and North America has been associated with increasing regulation and regularization of many areas of social life.

The cultural responses to earlier forms of industrial mechanization are too numerous and various to catalogue here, but we can give some examples for the sake of comparison with the era of proceduralization that we are now experiencing. In the first half of the 20th century, the distinction between elite and popular culture was still meaningful, and the two reacted differently to mass production and consumption. The avant-gardes of the time (elite in spite of their rejection of 19th-century art) reacted violently, but in contradictory ways. The Futurists famously embraced power technologies and even the technologies of war, whereas Dada and the surrealists were skeptical or even hostile to technological change. Popular culture, on the other hand, found that the new technologies offered new forms of entertainment and social interaction: amateur photography, the phonograph, radio, and film. Walter Benjamin (1968), of course, argued the case for film as a new aesthetic and political medium for the masses in the *Work of Art in an Age of Mechanical Reproduction*.

In the 1920s and 1930s, some popular and avant-garde filmmakers addressed directly the theme of the human cost of mechanization. Chaplin's *Modern Times* (1936) is a famous expression of the fear of subordinating humans to the machine. Its most memorable scene is the one in which the Little Tramp becomes mesmerized by his repetitive work on the assembly line, falls onto the conveyer, and is devoured and then regurgitated by the giant gears of the machine. In this case the popular filmmaker revolted against the mechanical, whereas the avant-garde Fernand Léger in the abstract *Ballet Mécanique* (1924) was fascinated, as was the Bauhaus artist Oskar Schlemmer in his theater piece, *Triadisches Ballet* (1922), in which human dancers dressed and performed like marionettes in 12 highly synchronized geometric pieces (Goldberg, 2001, pp. 111–113). Schlemmer seemed intrigued, rather than frightened, by the prospect of reducing his human dancers to movements that suggest the essence of the mechanical. Finally, Fritz Lang's *Metropolis* (1927), somewhere between the popular and the avant-garde, sent a mixed message of fascination with and horror at a mechanized utopia, in which workers are fed to the giant M-Machine, and a mad scientist wants to replace them all with robots. At the end of the film, the working and the elite classes are reconciled, suggesting that a mechanized future can accommodate human needs. The theme of the robot in film and theater from this period (from Capek's original robots[1] to Chaplin's robotically mesmerized Tramp to Schlemmer's mechanical dancers and finally Lang's lascivious female robot, Maria) is a small part of the culture's reaction to the mechanical, but it does indicate an understandable concern with the prospect of human assimilation to technology.

Robots appear in science-fiction films through the second half of the 20th century, ostensibly posing the same question about the nature of the human in an age of mechanization. We need only think of the most compelling robot of that whole period, HAL in *2001: A Space Odyssey* (1968), whose body is reduced to an ominous and all-seeing eye, to realize the extent to which the cultural understanding of the mechanical had changed. Between Maria in 1927 and HAL in 1968, the fully electronic computer had been developed and provided a new metaphor for mechanization. The physical, often geometric, representations of robotic movement were replaced by a fascination with artificial intelligence as a kind of mechanization of the mind. Gears were replaced with algorithms. The computer program is still a mechanism, but it is characterized by a flexibility that metal gears cannot achieve. When the astronaut enters the machine room in *2001* in order to switch HAL off, the depths of this machine are nothing like the one that swallows the Tramp in *Modern Times*. HAL consists of gleaming, symmetric rows of CPU and memory modules; what matters, what threatens the astronaut, is the invisible and hugely complex program that is racing around inside the physical circuits. The mechanization of the computer is so different from that of the traditional industrial machine that it has been given a new name: many authors today refer to *procedurality* rather than mechanization.

Ian Bogost (2007) wrote, for example: "Procedurality is the principal value of the computer, which creates meaning through the interaction of algorithms.... This ability to execute a series of rules fundamentally separates computers from other media" (p. x).

Bogost also spoke of media rather than machines in general here, because the computer now functions in our culture as a new medium or set of media forms.

Earlier media apparatuses, like other mechanical devices, could in fact execute a series of rules, but it was difficult to change the rules, because they were embodied in a physical mechanism. A photographic camera embodied the rules for taking one picture after another by opening its shutter, focusing the light on the film, and then closing again. Although that series of rules was just about all that a simple box camera could execute, an expensive analogue camera allowed for a large number of minor adjustments (to the focus and the shutter speed, for example). On the other hand, as computer specialists and popular writers have been pointing out since the 1950s, a fully programmable digital computer is capable of becoming infinitely many different machines—any machine that can represent and operate on discrete symbols. If we looked at the development of calculating machines from Hollerith's punch-card tabulators around 1900 to the UNIVAC and other electronic computers of the 1950s, we would see stages in the transition from mechanism to procedure. Whereas the tabulators that made IBM a major company in the 1930s and 1940s were fully mechanical, by 1944, the IBM Automatic Sequence Controlled Calculator or Mark 1 represented a transition to an electro-mechanical system of switches and relays. By the late 1940s computers were fully electronic, representing numbers and performing operations through the use of (still rather unreliable) vacuum tubes. Electronic construction permitted the crucial innovation that characterized the so-called Von Neumann machines: the program was stored in the machine along with the data. The program, the series of rules, could then be changed for each run of the machine, and this flexibility constituted and still constitutes procedurality as Bogost and others understand it today.

Just as we can identify a series of machines that describes the development from mechanical calculators to the first electronic computers and then to the powerful computers of today, we could also identify steps in the increasing proceduralization of human beings in 20th century society. The process of perfecting mass bureaucracy went on throughout this period: both in the formation of the welfare state and in its evil parodies, e.g., the meticulous National Socialist bookkeeping of the Holocaust and the bureaucracy of the Soviet Gulag. This process involved assigning people to categories (rudimentary database design) and applying increasingly sophisticated sets of rules to their social lives. The demands of the database and complexity of the rules eventually required computers for their management. The American Social Security system adopted computers in the early 1950s, converting its mountains of punched cards processed by mechanical tabulators into tapes processed by computer. If proceduralization was an ongoing trend throughout the century, it is hard to imagine that it could have penetrated so deeply into the grain of contemporary society without digital technology.

At one level, then, procedurality is the intensification of the mechanical introduced into social forms. This point seems crucial to a critic of digital culture such as Nicolas Carr, who in *The Shallows* (2010) equated the digital project of Google today to the industrial project of Frederick Winslow Taylor in the heyday of mechanization.

By breaking down every job into a sequence of small, discrete steps and then testing different ways of performing each one, Taylor created a set of precise instructions—an algorithm, we might say today—for how each worker should work. Taylor's system is still very much with us; it remains the ethic of industrial manufacturing. And now, thanks to the growing power that computer engineers and software coders wield over our intellectual lives, Taylor's ethic is beginning to govern the realm of the mind as well. Google's headquarters, in Mountain View, California—the Googleplex—is the Internet's high church, and the religion practiced inside its walls is Taylorism. Google, said its chief executive, Eric Schmidt, is "a company that's founded around the science of measurement," and it is striving to "systematize everything" it does. What Taylor did for the work of the hand, Google is doing for the work of the mind (Carr, p. x).

But Carr was eliding important differences between the mechanization of labor known as Taylorism and contemporary procedurality. This procedurality is not simply the extension of the practices of the mechanized assembly line to the realm of the digital. Bogost and others invoke computer games as defining examples of procedurality, and we can do the same to understand the differences between digital procedurality and the forms of mechanization it has superseded.

Computer Games and the Procedural

Espen Aarseth (2004) expressed a view that is common among those in games studies today, when he wrote that:

> The computer game is the art of simulation.... It is the dynamic aspect of the game that creates a consistent game world. Simulation is the hermeneutic Other of narrative, the alternative mode of discourse, bottom up and emergent where stories are top-down and preplanned. In simulations, knowledge and experience is created by the player's actions and strategies, rather than recreated by a writer or moviemaker. (p. 52)

Computer applications are sometimes, though not always, used to run simulations. Videogames in particular sometimes provide a coherent operating replica of some activity we know in the world, and when they do, they may be called simulations. Videogames are, however, always procedural: they are programmed to run through a series of steps (show graphics, play sounds, tally points) in response to the user's input. This is as true of MMORGs (Massively Multiplayer Online Roleplaying Games, also referred to as MMORPGs or MMOs) such as *World of Warcraft* as of single-player puzzle games. The wildly popular puzzle game *Tetris*, for example, is not a simulation. To say that it simulates falling blocks is misleading, for we learn nothing about how blocks fall in the world by playing *Tetris*. However, *Tetris* is procedural: its set of programmed rules is visualized as falling blocks and the moves the player makes to manipulate them. The player of *Tetris* works feverishly to get the blocks in the right places, like Chaplin's Tramp at the assembly line in *Modern Times*, except that in the case of *Tetris* this frenzied activity is experienced as play rather than alienating labor.

World of Warcraft and *Tetris* illustrate elements of procedurality that characterize our interaction not only with videogames but also with digital technologies in general: parameterization and event loops. In most games where players assume roles or take control of active figures, their ability to act and the results of the actions in the game are represented by a set of parameters. Discrete values indicate what kind of character you are playing (wizard, warrior, priest, and so on) and how powerful, healthy, or skilled you are. If you find food, your health points may rise. If in the course of combat, your health points dwindle to 0, you die.

The second element, the event loop, is less apparent, but no less important. The event loop, depicted in Figure 3.1, is basic to the procedural architecture not only of videogames, but also of interactive applications in general. The machine loops through a series of programmed actions, and at very short intervals checks to see whether the user has clicked the mouse, pressed a key, or moved a joystick. Applications built around one or more such event loops in this way are interactive, because they invite the user to intervene or interact. Although interactivity is a ubiquitous term in popular descriptions of digital media, proceduralists often criticize the term as vague. What seems important to them is not merely that the user interacts, but that she interacts with an algorithm, the code that lies beneath the surface of the application.

Videogames are particularly good at coupling the level of the code with the interface—what the player sees on the screen and what she does at the keyboard or game console. Videogames must effect a tight coupling, because the point is to engage the player intently and immediately in the play. Games have been characterized as "flow" experiences, because they promote a flowing or even addictive relationship between the player and the action (see, for example, Schell, 2008). The key to promoting flow is to insert the player so seamlessly into the event loop that she feels herself part of the procedure itself and wants the loop to continue indefinitely.

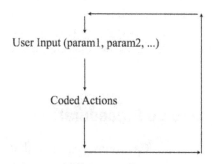

Figure 3.1: Procedural interactivity

Videogames are particularly good at coupling the level of the code with the interface: what the player sees on the screen and what she does at the keyboard or game console. Videogames must effect a tight coupling, because the point is to engage the player intently and immediately in the play. Games have been characterized as "flow" experiences, because they promote a flowing or even addictive relationship between the player and the action (see, for example, Schell, 2008). The key to promoting flow is to insert the player so seamlessly into the event loop that she feels herself part of the procedure itself and wants the loop to continue indefinitely.

Videogames have often been criticized for promoting violence in teenagers or for discouraging them from more productive activities and social relationships. Some media writers, among them James Gee (2007) and Steven Johnson (2006), have responded by contending that games actually promote positive social values or sharpen the players' minds. As Johnson put it,

> It's not what you're thinking about when you're playing a game, it's the way you're thinking that matters…. [G]ames force you to decide, to choose, to prioritize. All the intellectual benefits of gaming derive from this fundamental virtue, because learning how to think is ultimately about learning to make the right decisions: weighing evidence, analyzing situations, consulting your long-term goals, and then deciding. (pp. 40–41)

Although Johnson did not put it this way, videogames stimulate decision-making through procedures. The game absorbs the player into its event loops. The procedures constitute the situations that require the player's decisions, and the decisions themselves are always expressed as inputs that are processed in the loops. Videogames have a specific educational function, then: they train us to take part in the procedural structures of contemporary computer systems. Such training is needed today because of the ubiquity of those systems.

Learning to Be Procedural

In *Synthetic Worlds* (2005), Edward Castronova argued that the virtual economies of MMOs share many of the characteristics of economies in the material world. Second Life (secondlife.com), for example, is an MMO with tens of thousands of regular participants, a virtual world that appears as a three-dimensional environment that their avatars can explore and in which they own land and create 3D models. In exchange for a monthly subscription fee, they receive a plot of land and a monthly stipend of Linden dollars. This virtual currency underlies the economy of Second Life, because it can be used to buy land, houses, clothes, or the services of other avatars. Linden dollars can also be changed back into U.S. dollars. For Castronova, the markets in MMOs such as Second Life act like the markets of our first life in the material world. What Castronova does not explicitly consider is that the reverse is also true. Our everyday exchange of goods and services is conducted increasingly through the procedurality of digital environments and virtual worlds. Online shopping, for example, becomes more economically important each year. In the United States, Black Friday, the most rapacious shopping day in the malls of America, is now followed by Cyber Monday, the day of the greatest online sales. Customers have become accustomed to inserting themselves into digital event loops for the online purchase of films, books, clothes, airline tickets, hotel rooms, and computers themselves. A recent Pew survey suggested that more than half of all American adults have shopped online and nearly half bank online (Zickuhr, 2010).

Our bank accounts are in fact tiny portals into an enormous economic network that in itself already constitutes a virtual world. In creating our accounts, we fill in our profile parameters and check preferences as if we are setting up avatars in a role-playing game. In examining our balances and paying bills, we manipulate the interface as in a simulation game. Our bank accounts, indeed the whole economy, have become a giant simulation—we can hardly help but think of William Gibson's definition of cyberspace—running on "the banks of every computer in the human system. Unthinkable complexity" (1984, p. 51). It was precisely the fact that the economy is an unthinkably complex simulation that encouraged the manipulation of monetary abstractions such as "credit-default swaps" and "mortgage-backed securities" and very nearly crashed the whole system in 2008.

Steven Johnson was right, then, that videogames teach young people skills they will need in order to participate in contemporary society. Even older adults, who were not raised on such games, have become adept at online banking and shopping. In one sense, participation in today's digital economy is more intricate, but not essentially different from the demands of 20th century economic and bureaucratic life. But because the procedurality of the computer is so flexible and responsive, our relationship to technology is now subtler and more intimate than in the age of mass industrialization. Along with our economic activities, social life is becoming increasingly proceduralized. Users of all ages seem now to enjoy the proceduralization of entertainment and forms of social communication.

iTunes is perhaps the leading example of proceduralized entertainment: a site where hundreds of millions of users purchase and organize songs, movies, and television shows in increasingly elaborate ways. iTunes invites us to parameterize and program our playlists or to use the Genius feature to create them automatically. We share our lists and preferences with others through Ping or on Facebook or Twitter, so that entertainment and social communication flow easily into one another. The point of iTunes, YouTube, Facebook, Flickr, Twitter, and numerous other such procedural systems is to keep us flowing along: moving from listening to watching, sharing, and adjusting our parameters.

The purpose of the interaction design for these systems is to conduct the user subtly and effortlessly into the event loop. Steve Jobs and his design team at Apple were noted for artifacts that present the procedural as "magic"—an aesthetic in which the underlying code remained invisible beneath a seamless user interface. Apple under Jobs practiced a kind of modernist design in which form followed function and still managed to make the function seem playful and surprising. Apple's iPhones and iPads would in fact only appear magical to a highly technologically literate audience that already regarded procedurality as natural. Good digital design today encourages its users to proceduralize their behavior in order to enter into the interactions, and a large portion of those in developed countries have now accepted this as the path to participation in digital media culture.

In addition to new procedural media forms such as Facebook and iTunes, the traditional media of film and television increasingly offer procedurality as a theme or represent it as a technique. In the case of television, the term "police procedural" was coined as early as the 1950s to describe formulaic crime dramas built around the procedures of police work (Prial, 2005). Until the 1980s almost all television dramas and sitcoms were procedural in another sense too: each episode followed an established formula, and the characters and conflicts were reset at the beginning of each new episode. In other words, each episode looped back to the starting condition, so that viewers could watch reruns in any particular order or skip episodes without getting lost. This was television in the early era of computers, when programs processed data repetitively, but were not interactive, and many television series are still procedural in this sense.

More recently, action-adventure films have become increasingly adept, if not obsessed, with procedural scenarios, in which good or evil teams of specialists carry out heists or paramilitary actions with a programmed precision that often defies nature. The *Mission Impossible* series is an ongoing example, with each new film topping the last in split-second timing and choreographed violence. *Ocean's Eleven* and *Ocean's Twelve* introduce self-referential humor and play with a lighter touch, but still revel in the procedural. Heist films constitute a venerable Hollywood genre and have always depended on coordinating the human participants in a procedural plan. But recent examples always seem to include a computer geek who coordinates the action by blocking security systems, taking down power grids, and warning his team of approaching enemies by tapping into surveillance cameras. Much of Christopher Nolan's heist film *Inception* (2010) takes place in a computer-induced collective dream in which different time scales in different levels of dreaming must all be coordinated to the second. As we noted, Nolan's earlier *Dark Knight* gave us a character who purported to be the enemy of the procedural. In these two films, then, Nolan has explored both our cultural fascination and ambivalence with proceduralization.

Meanwhile, digital media subcultures have developed linear (noninteractive) graphic or video forms that also celebrate procedurality. The socalled "demoscene" shows its fascination with the procedural by programming virtuoso computer-graphic displays that run in real-time. Demoscene contests such as Breakpoint (breakpoint.untergrund.net) are hacking festivals that foreground the ability of programmers to make visual displays exclusively through the code. There are no captured images, as is the case with DVD videos, YouTube videos, or any of the commercial forms (en. wikipedia.org/wiki/Demoscene). Often these hackers also work with extreme space limitations: the whole program may be limited to 65K bytes or even 8K bytes. The aesthetics of these demoscene pieces is often quite conventional: the creativity is measured not by the programmer's ability to test the visual conventions of digital animation, but rather by the ability to fashion complex moving images from pure procedurality. Procedural animation now occurs in mainstream games and films as well. The elaborate videogame *Spore*, for example, makes considerable use of this procedural animation (en.wikipedia.org/wiki/Spore_(2008_video_game)), and the background music for *Spore* created by Brian Eno was in part procedural as well, generated by algorithm rather than recorded and digitized (Thiessen, 2008).

Procedure and Performance

Interactivity is a term that digital media writers have often criticized but cannot seem to dispense with. Among these writers, especially in game studies, the term is used to describe the relationship between the player and the program or game itself. The player inserts herself into the event loop and issues some action, to which the program responds, to which she in turn responds, to which the program again responds, and so on. In contrast to games studies, communication studies has a different understanding of interactive media. For example, in *Communication Theory: Media, Technology, Society* (2005), David Holmes made it clear that media interactions occur between human subjects. By this definition, the telephone is interactive, but broadcast radio is not. In the digital realm, email is an interactive media form, as is Twitter. The single-player videogame is not interactive because it is not intersubjective: the machine does not count as a subject. On the other hand, the multiple-player game is interactive, because the players co-opt the game as a medium or channel for interpersonal communication.

These two definitions of interactivity define two distinctly different ways in which users or player may perform in their encounters with digital media. In *Perform or Else* (2001), Jon McKenzie pointed out that, among its several meanings, our term "performance" can refer to a technical ability to achieve a goal, or it can refer to the act of presenting ourselves in a certain way to an audience. When someone plays a single-player videogame, she is performing by testing herself against the program. This sense of performance as a measure of achievement corresponds to the sense of interactivity as human subject against machine. The other sense of performance, as acting or presenting ourselves, is appropriate to the other definition of interactivity. In a chat, a blog, or a Twitter post, we present ourselves to an audience, though perhaps an audience of one. Our performance is for people, although it is mediated through digital technology.

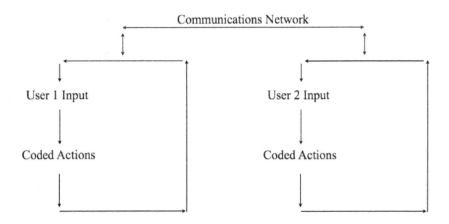

Figure 3.2: Computer-mediated intersubjectivity

The single-player game is one in which we have a single loop, as was shown in Figure 3.1 above. So is a word processor or iTunes. Social media, MMOs, and chat programs can be represented with a double loop (Figure 3.2). Each user loops through her client program on her computer, while the clients themselves communicate with each other, either synchronously or asynchronously, through the network and usually through a server. The digital system mediates the performance of each human user for the other.

These performances are all still procedural, although they are more flexible than any single-user program can be, for the obvious reason that they involve human performers on both ends, or in fact numerous performers in the case of MMOs and other group media. Both forms of interactivity involve performing in and through event loops.

For decades, performance studies has examined rituals as cultural performances occurring in an arena that is neither the same as everyday life nor like the theater (see, for example, Schechner, 2006). Rituals are perhaps always procedural, in that they require the participants to follow a series of defined steps toward a goal. Unlike classic rituals of coming of age or marriage, however, social media, MMOs, and other interactive digital media are everyday practices. If they once seemed to take place in the special world of cyberspace, that is no longer the case because of the ubiquity of such media forms (on desktops, laptops, tablets, smartphones) and the frequency of their use. Facebook and other social networking sites offer repetitive communication rituals that have massive appeal in every developed country. One important difference with the rituals of social media is that the procedures are defined not only by cultural agreements, but also by the code itself. The code dictates how much you can communicate, in what forms (text, images, videos), and to whom (friends or larger user communities). Facebook is the most popular such coded ritual, or indeed an expanding set of such rituals: setting up your Facebook page, adding images, writing on friends' walls, sending a message to a friend, and so on. Each Facebook application or game adds a different (though not radically or disruptively different) ritual activity. New Facebook applications can be successful only if they provide enough structure to render the ritual intelligible while still allowing enough scope to communicate something satisfying to fellow users.

The success of social media forms in general seems to depend on constructing procedures that are both constraining and flexible. The wall in Facebook makes it easy to write messages to all your friends at once and to read theirs. The ability to change your profile, upload images, and subscribe to and share games gives users scope for playful construction of their identity within procedural constraints. In all social media, the user is still inserting herself into the code, becoming part of a procedure. But in each application she does so in a different way, like an actor performing a different script. The capacious, multimedia pages of Facebook, which promise an ever increasing number of paths for connecting with friends, define a different set of possibilities from the rapid, laconic presentation of self in Twitter.

Many have studied the way in which various social media facilitate the presentation of self in the peculiar context of online media (see, for example, boyd & Ellison, 2007; Papacharissi, 2009, 2010). Such studies often look back to the work of Erving Goffman's *Presentation of Self in Everyday Life* (1959). Goffman examined how people in his contemporary society (the 1950s) performed their identities in face-to-face situations, and many investigating social media today are trying to determine whether Goffman's categories can apply to online interactions. The key difference is that Goffman described practices that are governed entirely by social rules, whereas the digital definition of self is produced through the interplay of social practices and procedures that the user cannot easily change.

Alternate Reality Games and the Triumph of Procedurality

Social media are no longer limited to the World Wide Web; smartphones and tablets allow users to bring their communication rituals with them as they go through their day. Alternate Reality Games (ARGs), which first appeared about 10 years ago and therefore at about the same time as social media themselves, offer another avenue for extending procedurality into the world. Early examples, such as *The Beast* and *I Love Bees*, were elaborate multimedia conspiracy games. The players were invited surreptitiously into the games through clues scattered in Web sites or movie trailers. A so-called "rabbit hole" for *The Beast* was a credit in trailers and posters for film *A.I.*, listing Jeanine Salla as "Sentient Machine Therapist." This sent some tiny fraction of the viewers to look for a Jeanine Salla Web site, and they in turn enlisted friends to help them unravel the game's mysteries. The players' actions were relatively restricted: they could not create new material, but they were encouraged to comb the Web as well as physical locations to find clues and solve puzzles. The players worked in groups, and their performance consisted of the collective reading of a story written both online and in the physical world. More recent games, such as those devised by Jane McGonigal, expand the field of performance for their players. In *World Without Oil* (2007), there was no mystery to uncover, but rather an evolving economic and social scenario to respond to. In the game's scenario an oil shortage was causing drastic and repeated price shocks, and the players were to respond by describing in blogs and videos the impact on their community. Some players chose to perform their responses in the world—by bicycling to work or adjusting other activities as if gasoline were actually in short supply. McGonigal offered such ARGs as examples of a new style of activism that can effect social change (McGonigal, 2011).

There was also a promotional ARG for *The Dark Knight* entitled *Why So Serious?* which preceded the release of the film by 14 months. The seed in this case was an email sent to thousands of potential players, inviting them to visit a Web site, which in turn led them to a bakery and a cake containing a cell phone. Following a series of clues brought persistent players in several cities to secret screenings of the first minutes of the film. The screenings suggested that, through their actions in solving the puzzles, the ARG players had become accomplices to the Joker's crime (Rose, 2011). By staging this ARG, which was in the end an elaborate mechanism for drawing a select audience to a preview, Nolan and his collaborators made *The Dark Knight* into a cross-media project. They also confirmed and elaborated a theme in *The Dark Knight* itself, foregrounding our culture's fascination, and at the same time its ambivalence, with the procedural. It is not surprising that a Hollywood film would seek to address our reaction to digital proceduralization. Both as an industry and a cultural form, cinema feels threatened by the rise of digital media—above all, by videogames but also by social media and pervasive games as well.

Those who work in new digital media forms are, on the other hand, understandably more sanguine. In *Reality Is Broken* (2011), Jane McGonigal argued earnestly that playing collaborative videogames, including but not limited to ARGs, can change the world:

> Very big games represent the future of collaboration. They are, quite simply, the best hope we have for solving the most complex problems of our time. They are giving more people than ever before in human history the opportunity to do work that really matters, and to participate directly in changing the whole world. (location 5662, Kindle edition)

McGonigal believes that the procedurality of videogames can be harnessed as a collective intelligence and brought to bear on social and economic problems. The success of her vision depends on the willingness of tens of millions of participants to insert themselves into the event loops of collaborative games. It is a vision in which procedurality triumphs as a cultural practice.

Note

1. Czech writer Karel Capek introduced the word "robot" in his 1920 play *R. U. R. (Rossum's Universal Robots)*.

References

Aarseth, E. (2004). Genre trouble: Narrativism and the art of simulation. In P. Harrigan & N. Wardrip-Fruin (Eds.), *First person: New media as story, performance, and game* (pp. 45–55). Cambridge, MA: MIT Press.

Benjamin, W. (1968). The work of art in the age of mechanical reproduction. In H. Arendt (Ed.) (H. Zohn, Trans.). *Illuminations: Essays and reflections* (pp. 217–251). New York, NY: Schocken Books.

Bogost, I. (2007). *Persuasive games: The expressive power of videogames.* Cambridge, MA: MIT Press.

boyd, d., & Ellison, N. B. (2007). Social network sites: Definition, history, and scholarship. *Journal of Computer-Mediated Communication, 13*(1), 210–230.

Carr, N. (2010). *The shallows: What the Internet is doing to our brains.* New York, NY: W.W. Norton.

Castronova, E. (2005). *Synthetic worlds: The business and culture of online games.* Chicago, IL: The University of Chicago Press.

Gee, J. P. (2007). *What video games have to teach us about learning and literacy* (2nd ed.). New York, NY: Palgrave Macmillan.

Gibson, W. (1984). *Neuromancer.* New York, NY: Ace Books.

Goffman, E. (1959). *The presentation of self in everyday life.* New York, NY: Anchor.

Goldberg, R. (2001). *Performance art: From futurism to the present.* London, England: Thames and Hudson.

Holmes, D. (2005). *Communication theory: Media, technology and society.* London, England: Sage.

Johnson, S. (2006). *Everything bad is good for you: How today's popular culture is actually making us smarter.* New York, NY: Riverhead.

McGonigal, J. (2011). *Reality is broken: Why games make us better and how they can change the world.* New York, NY: Penguin.

McKenzie, J. (2001). *Perform or else: From discipline to performance.* London, England: Routledge.

Mumford, L. (1934/2010). *Technics and civilization.* Chicago, IL: The University of Chicago Press.

Nolan C., Roven, C., & Thomas, E. (Producers), & Nolan, C. (Director). (2008). *The Dark Knight* [DVD]. United States: Warner Brothers.

Papacharissi, Z. (2009). The virtual geographies of social networks: A comparative analysis of Facebook, Linkedin and Asmallworld. *New Media & Society, 11*(1/2), 199–220.

Papacharissi, Z. (Ed.). (2010). *A networked self: Identity, community, and culture on social network sites.* New York, NY: Routledge.

Prial, F. J. (2005, July 9). Why readers keep returning to the 87th precinct. *The New York Times.* Retrieved February 28, 2012, from http://www.nytimes.com/2005/07/09/books/09mcba.html?pagewanted=all

Rose, F. (2011). *The art of immersion: How the digital generation is remaking Hollywood, Madison Avenue, and the way we tell stories.* New York, NY: W.W. Norton.

Schechner, R. (2006). *Performance studies: An introduction* (2nd ed.). New York, NY: Routledge.

Schell, J. (2008). *The art of game design: A book of lenses.* Burlington, MA: Morgan Kaufmann.

Thiessen, B. (2008, September 16). *Brian Eno scores spore.* Retrieved February 27, 2012, from http://exclaim.ca/News/brian_eno_scores_spore

Zickuhr, K. (2010, December 16). *Generations 2010.* Retrieved February 27, 2012, from http://pewinternet.org/Reports/2010/Generations-2010.aspx

Technobiography: Industry, Agency and the Networked Body

Eric Freedman

In this chapter I explore the "life technobiographic"—the life written through technology and approximated by a data trail (a rich record of subjective presence)—and consider the manner in which certain technological dependencies may be driven by private sector research. Technobiography implies a particular form of authorship and calls for an understanding of how the self is situated within social relations that inherently involve engaging with information technologies. I am less concerned with story (the technobiographic artifact) than with narrative (the technobiographic process). By speaking of the life technobiographic, I am articulating the manner in which subjectivity is transformed by new technologies, for those very same technologies affect how we translate everyday events into a knowledge system. Subjectivities, knowledges, and technologies are always situated; they are all regularly appropriated, rehistoricized, and read anew. By linking technology to biography, I acknowledge the fundamental power of enunciation and hold onto the dual possibility of acting out and being acted upon by metadata.

To approximate the life technobiographic, in the most obvious and commonplace way, we might begin by Googling ourselves, gathering every self-inflected node. This rather abstract portrait lets us see how diffuse we have become, determined as we are to infiltrate networked space. Here, diffusion is a sign of success. We might move on to examine our smart devices and consider the personal preferences we have used to imprint ourselves on them. We might study our nuanced interactions with the interfaces we encounter every day—recording, communicating, producing, consuming, and transacting. Still, taken together, these tactics, this sum total of signs, will not speak the technobiographic subject. There is more work to do, and these signs are merely symptomatic of a process of mutual inflection. I have argued (Freedman, 2011) that in an age of pervasive computing we must read biography against the dominant forms of technological projection and consider how technology too is practiced. This chapter follows the basic contours of a project undertaken by Michel Foucault (1970) in *The Order of Things*, his most fervent critique of the constitutive limits of discourse (which, he contended, is guided by historically situated epistemes). Foucault's study led him to ask why we persistently turn to certain modalities of order to make sense of everyday life; proceeding from this query, my goal is to highlight several key pressure points between personal agency and industrial design, as autobiography is a dynamic process commonly channeled through static sign systems.

In this chapter, I isolate several contemporary modalities and draw attention to the discourse to reveal the selective deployment of new technologies. Technology is regularly positioned as a technobiographic agent, where subjectivity may be understood as a series of encounters with technology; in essence, it is written through them, and recorded and shaped by them. The formation of subjectivity suggests a movement through a technologically inflected landscape at once psychological and geographical, and we can analyze this psychogeography to temporally unfold the subject.

The focus of this analysis is the contouring of the technobiographic subject through multiple industrial frameworks, especially its networkability. This study is firmly grounded in an institutional analysis and illustrates the more nuanced mechanisms of political economy. The social, individual, cultural, technological, and industrial forces that shape media are often overlooked when they are embedded in intimate material practices, and are difficult to analyze when those practices are distributed. Even so, because cultural practices commonly rely on material goods and services, we must consider the varied operational limits of the citizen-consumer in social, political, and economic discourse. The citizen-consumer performs through commodities, choosing and often exceeding them, yet this subject always does so in relation to both a local and a global field and as part of a complex interaction between the two processes of consumption and citizenship. Marcuse (1964) warned: "In advanced capitalism, technical rationality is embodied, in spite of its irrational use, in the productive apparatus" (p. 22), and cautioned about the loss of viable space for transcending historical practice. The determinations governing the networked body may not be strictly technological or economic, but they do exist, channeled into general formations about the technological imaginary, which are used to argue for the particular form that fabricated (or industrialized) network culture may take. The periodic yet requisite fusion of intellectual energies in the technology sector is mostly forged from economic imperatives. Unfortunately, the institutional rigidity that governs corporate research and development largely divorces innovation from any humanist criteria, although innovation may find its inspiration there. Nevertheless, industrial control and mediation have decided sociological consequences.

Industry

With the launch of the iPhone 4S in October 2011, came Siri, Apple's voice-controlled artificial intelligence assistant. Mapped onto the iPhone's home button, it reads as a natural extension of Apple's mobile platform. The app is commonly touch and voice activated, but can be launched with a "raise to speak" feature using an internal infrared proximity sensor that tracks the phone's distance from the body. In both modes, the app encourages new forms of multisensory engagement, making the device a seamless evolution of previous iterations. Acutely aware of these emergent forms of physical and emotional intimacy (responding to tactile and verbal cues, and fulfilling the end user's wants and needs), Apple has humanized the device: "It's like you're having a conversation with your iPhone" (n.d.). Yet these personal attachments are structured by underlying algorithms linking semantic intent to information retrieval (in effect translating natural language into data acquisition) and are the product of years of defense-sponsored research on speech recognition and artificial intelligence at Menlo Park, California-based SRI International. The technology was developed as part of the CALO (Cognitive Assistant that Learns and Organizes) project, an initiative launched by SRI in 2003 with $22 million in startup funds from the Defense Advanced Research Projects Agency (DARPA). In 2007, SRI created the subsidiary Siri, Inc. and secured additional financing from several venture capital firms in order to begin commercializing its work. Apple acquired the startup in April 2010 for an estimated $200 million (Panzarino, 2011). The high price tag is justified by the research grounding Siri, and the complex set of artificial intelligence insights running behind the app: the chain of machine-learning, natural-language processing, and Web search algorithms that drive each consumer query. With a mobile device such as the iPhone, these algorithms can be tethered to the contextual awareness of GPS location reading and end-user preferences to transform the search engine into a *do* engine. Siri has, in effect, transformed the iPhone into a body with a spatial awareness and a sense of intent.

As a non-profit research and development center doing contract research for the government and other clients, SRI International has focused on the conceptual and practical relations between science, technology, and communications since its founding as the Stanford Research Institute in 1946. A notable event in the institute's history was a public lecture by Douglas Engelbart on December 9, 1968 (Engelbart & English, 1968). He and 17 research colleagues from the Augmentation Research Center presented a live public demonstration of the online system, NLS, which they had been working on since 1962. Engelbart's tech-laden lecture was the public debut of many innovations, including the computer mouse, hypertext, object addressing and dynamic file linking, and shared screen collaboration involving two persons at different sites communicating over a network with an audio-visual interface. The session featured the computer-based, interactive, multiconsole display system being developed at SRI under the sponsorship of ARPA, NASA, and RADC, and was readily situated as part of a contemporaneous research agenda that suggested interactive computer aids could augment intellectual capability.

During his address, Engelbart took what at first seemed a rather humorous digression: "Let me go to a file that I prepared just after my wife called me and said, 'On the way home will you do a little shopping for me?'" But he proceeded by readily contextualizing the conversational fragment: he organized his shopping list into a series of numbered statements, and illustrated how the system can reorganize the terms according to the logic of commerce—the layout of the grocery store—as he grouped the list's produce items. He also visualized his to do list as a line diagram that crudely traced his route home, with stops at the library, the drugstore, the market, and other destinations; the data set became a task-oriented vector that itself is a formal abstraction of agency within a personalized geography.

Engelbart's demonstration can be situated as a necessary moment in the development of computer culture, which has been dependent on three interrelated technological pursuits: artificial intelligence, interface design (and concomitant computer design), and virtual and augmented reality. Ubiquitous computing, the sum total of these developments, is often positioned as a third wave of computing, after mainframes and the era of personal computing, a moment when, according to Weiser (1991), technologies "weave themselves into the fabric of everyday life until they are indistinguishable from it" (p. 94), and take the data out of information, leaving behind the ability to act.

SRI is not simply engaged in sponsored research, but also licenses its intellectual properties, bringing its innovations to the marketplace; the company has repeatedly created its own new corporate ventures with funding from investment firms, and has successfully monetized many of its proprietary technologies. A primary challenge in the commercial deployment of cognitive assistants is instilling trust in end users; it seems this tension has universal significance, because we might also understand it as necessary to any successful private-to-public-sector migration. As corporations launch new technologies, they follow the doctrines of commodity fetishism, emptying their mass-produced goods of the meaning of their production (the context in which they were produced and the labor that created them) to foster new and deeply personal relationships to their commodities; in essence, as years of research and development are channeled into fixed material relations, the initial impulse, the query, that sparked the intellectual endeavor must be willfully erased or recontoured.

Looking back at Siri's pre-history, the challenge to the CALO team was explaining the core architecture of its cognitive agent, the belief—desire—intention (BDI) framework that structures the mental state of the agent and plays an important role in determining its goal-oriented behaviors (Rao & Georgeff, 1991). The team determined that providing transparency into the system's reasoning and execution was key to establishing trust, to aligning the mental states of agent and consumer, both in their latent and active relations (McGuinness, Glass, Wolverton, & Pinheiro da Silva, 2007). The goal was to perfect the symbiotic attachments between the software and the user, and transparency seemed to be a necessary foundation for what would become a relatively calm technology. Translating the lessons of transparent BDI-based task processing into popular terms, Apple suggests, "Siri not only understands what you say, it's smart enough to know what you mean. ...And Siri is proactive, so it will question you until it finds what you're looking for" (n.d.).

This act of translation, this modeling of the commodity fetish, is not insignificant. To theorize hardware in the context of political economy may seem an obvious gesture, given hardware's essential materiality. To theorize software, network, or operating system may seem a purely academic exercise, but if the goal is to understand the ideological trappings of new technologies, we must be attentive to the architectures of both hardware and software. Just as ideology is a figurative projection, so too are the immaterial frameworks driving material devices. One difficulty in deciphering these systems is that in today's culture industry, the cultural and the industrial are coterminous, and the reverse engineering of algorithmic manipulation demands that programmers understand what end users are doing (Galloway, 2006). At the same time, consumers are acutely aware of the fetishistic logic framing their engagement with smart devices—that these relations are governed by interrelated forms of technical transcoding. Siri's allure is that she (and Siri is by default a "she" in the United States) is a multilingual answer agent, interpreting human speech, mediating between the language of protocols, and negotiating between applications and service providers. Unlike corporate answer agents, such as Anna (IKEA's online virtual assistant), her knowledge base is not readily circumscribed, nor is it driven by an obvious imperative.

Yet Siri still embodies the principles of postindustrial capital; as a body at once localized (at our fingertips) and dispersed (comingled with the data cloud), she offers up the fantasy of a body without organs (Deleuze & Guattari, 1987). This desiring (search) engine is part of a productivity suite, and as such runs antithetical to any experimental and potentially subversive practice that might interfere with the structures and desires of capitalism; Siri may invoke the agentive concept of "becoming," but her value is based on the reliability (the completeness) of her executable code and the certainty of its tasks and outcomes. The indeterminacy of her application, which speaks to the user's desire, is matched by the determinacy of her innate mechanics, which speaks to the commercially driven mandates of programming and design.

Agency

The general trend toward seamless mobility heralded in the research and development of new technologies is one that aims to integrate multiple feature-rich media devices and operating platforms across disparate yet commonly traversed spaces—the home, the car and the office. This trend is part of a larger projection of the future of liquid media—taking media and shaping it to the various circumstances that people find themselves in—that also wants to embroil the subject in the technology.[1] New media industries have adopted the rhetoric of engagement and empowerment, and are mobilizing consumers by drafting biographical practices that can be attached to individual authors. The aim is to create new media frameworks that replicate subjectivity (by emphasizing boundless possibility and transformation) and merge the lived context with an apparatus of production, fostering the development of technobiographies that write and rewrite the self through the postindustrial logic of new media.[2] Responsive technologies seem to situate end users as unique social actors, as inscribed data (but not governing code) accumulates and becomes symptomatic of our presence. New technologies may seem to operate freely to the extent that they act intuitively, but their intuition is by design; it is inherently the result of a script written by developers. As we become conscious of the possibilities for remapping technology, we must not overlook the limits of our own subjectivity, which is itself the product of an unseen script.

The technobiographic subject is constructed through multiple interrelated frameworks. It may be useful, as a start, to outline the actions that I believe are central to the life technobiographic: anthropomorphizing and humanizing technology, fostering dependencies with responsive technologies, using autobiography as a signature content referent, and helping individuals put autobiography into practice (Freedman, 2011). These actions are given form in a vast array of institutional spaces, narrated by each institution's respective and specialized discourse. We see them given form in advertising, industry, and education; they are militarized and often politicized, and comingled with pre-existing clinical and professional practices. Within these institutional rubrics, the technobiographic subject has been both celebrated and demonized. Skirting these valuations and variances for the moment, in the most general sense the technobiographic subject may be approximated by examining its systemic attachments to technology. We see this in the life of smart objects that record our personal preferences, and in the contouring of smart objects whose interfaces and intelligences have been carefully calibrated with the human subject in mind, making such technologies seem intuitive and responsive. And we see this in the type of fluidity we expect in our engagements with new technologies, a feature we demand rather habitually, regardless of the context (Freedman, 2011).

Although I am critical of blind media effects discourse, I turn to advertising to illustrate how technology is regularly positioned as a technobiographic agent. Technology is consistently anthropomorphized and humanized throughout Apple's multigenerational advertising legacy, beginning with its *1984* ad, a tale of human resistance in a PC-laden Orwellian society. Several decades later, perhaps as a lighthearted reprisal of the troubling man/machine dyad, a 2002 spot for Apple's redesigned iMac (a flat-screen monitor mounted to a semispherical CPU by a swivel arm) features a man standing outside a store window as he is taunted by an iMac that mimics his every movement. Anthropomorphism segues into humanism in Apple's more recent *Get a Mac* campaign. Opening with the now-familiar greeting, "Hello, I'm a Mac," the spots use actors to play the competing architectures of Mac and PC. Pushing beyond the two platforms, the campaign's later spots expand the human chain. *Network* is designed to highlight Mac's compatibility and casts a Japanese woman[3] as the embodiment of a digital camera. In an effort to familiarize and demystify, and to insert technology into active citizenship, these ads efface technology altogether, and cast it only as a series of human equivalences.

Pushing human agency toward technocentric certitude—positioning technologies as causal influences in the social trajectories of individuals and their developmental pathways—the iPhone has been repeatedly framed as a biofeedback device. Its first clinical turn was the 2006 Nike-Apple partnership, pairing the phone with the Nike+, a wireless in-shoe accelerometer recording runners' workout statistics. When Apple unveiled details about its iPhone 3.0 software in March 2009, the company also announced a new class of third-party peripheral development that furthered this biometric trend: iPhone-compliant medical devices (a blood pressure cuff and a Johnson & Johnson LifeScan blood glucose meter). These accessories and their respective application suites enable the recording, charting, and transmission of physiological data, connecting patients to their bodies and to their health-care providers. The iOS blood pressure monitor is one of several intelligent devices manufactured by Withings (which also offers a WiFi body scale, and a baby scale and monitor) compatible with Web-based records services such as Microsoft HealthVault; taken together, these devices form a cohesive network of family analytics, turning surveillance into a discrete metric. They are understood as productive diagnostic instruments rather than as repressive apparatuses.

The iPhone (and more recently the iPad) has been positioned as a fundamental aspect of well-being, yet such dual positioning as communication tool and personal technobiographic agent is not unique. Microsoft Research introduced the SenseCam in 1999, a wearable digital camera designed to take photos passively, without user intervention. The device has been used as a tool to complement the restoration of autobiographical memory in clinical trials to treat patients with a broad range of memory disorders. SenseCam studies have highlighted the relative importance of images to the distinct memory processes of knowing (an act of pure inference) and remembering (the production of mental re-experience). But from a metacritical vantage point, the patient's dependence on the image is accompanied by a dependence on the technology itself; both are instrumental to restoring otherwise inhibited mental processes. The visual artifact seems to be of a first order, perhaps because of the immediacy of its affect, yet it is connected to the unrepresentable process of remembering; it is an aid to autobiographical memory but it is not autobiography proper. Similarly, the SenseCam device seems most readily aligned with its technological antecedents (it is, after all, a camera), but it is more appropriately aligned with the unrepresentable process of cognition. Most external memory aids improve prospective memory function, restoring agency to impaired subjects in direct ways (helping them to remember to keep appointments, take medications, and so forth), and very few work on the level of second order agency, to restore autobiographical memory and contribute to self-identity and self-concept (Berry et al., 2007). Most subjects in the SenseCam studies are acutely aware that the technology has played a critical role in their everyday lives, even if it can only construct an autobiographical arc commencing from the date of acquisition. It is difficult to frame this therapeutic intervention as yet another instance when technology is readily aligned with a certain degree of ideological fetishism, another instance of misrecognition derived from a fundamentally empirical or technical set of relations (Galloway, 2006).

Distillation is an ontological process discussed repeatedly in critical examinations of photographic practice, and the SenseCam fosters such agitation and sublimation; it reorganizes lived visual terrain and turns it into an abstract network of signs. Sontag (1990) suggested that with the passage of time, a photograph loses its specificity and becomes a purely aesthetic object. Barthes (1981) suggested that a photograph can do little more than confirm the existence of an object at some other time. But distillation is also an enterprise and therefore speaks to desire, however real or constructed the need. The quotidian and rhythmic (snapped at timed intervals) nature of SenseCam images suggests reflective yet forward movement rather than deeply nostalgic longing. Whether or not the photograph can stand in as a memory image, it can serve as a practical mnemonic aid. We want the image to be an active part of the narrative of memory, even if this desire is a fleeting impulse driving us to capture an image that will only be forgotten. Although photography is part of commodity culture and may be complicit with traditional economies of circulation under the laws of capitalism, it is nonetheless a highly charged psychic endeavor and, as participants in this exchange, we seem to want to insist on the authentic nature of its rituals and the ceremonies, even as we acknowledge our rampant idealism. Bourdieu (1990) suggested that the psychological explanation of photographic engagement is insufficient; to account for photography as a practice and to understand its rhythms, we must also attend to sociological accounts of production, distribution, and reception. To admit that the form of a medium is also a product of technical possibility and material and economic circumstance is to acknowledge the limits of individual agency, even in the most intimate and therapeutic exchanges.

Like the iPhone, the SenseCam is part of a larger history of the evolution of an apparatus, in this case, of the varied technological practices associated with life-logging (the use of wearable technologies to create person digital archives). These practices bridge the work of research laboratories with the commercial deployment of parallel technologies, and they bridge multiple media forms and object lessons—from the SenseCam and its images, to the Nike+ Sensor and its personal running metrics. Of these, the iPhone marks the most fervent adoption of the tools of industrial development, and the most seamless integration of digital mechanics and everyday life.

All of these practices reveal the evolving contours of technobiography; they speak to the degree to which the body is a network, experience can be quantified, and life can be lived through data. Because of their diversity, these applications also foster a greater understanding of the nature of auto-biographical knowledge, and the relative utility of autobiographical ex-change. Imperatives to record lived experience are often matched by imperatives to make the personal data trail public and to connect with others. In this scenario, the community (of other bodies) functions as a yardstick for our wellness, and the online social network becomes part of a corporate su-pertext extending the life of material goods. By touting the healthy rewards of its proprietary online social network, Nike, for example, encourages its consumers to share their personal data networks, and broadens the reach of its sensors. As a bridged network of signs, the data trails of Nike runners re-veal their interrelated states of well being, to the extent that an affective di-mension may be aligned with a physiological signature that is itself a technological projection recorded by a biometric device and displayed as an iconographic signifier on the Web. In popular media reports, Nike executive Stefan Olander referred to the data as a form of currency, as a sign of a con-sumer's energy (Greene, 2008), a concept the company took even further with its Nike+ FuelBand (and the associated NikeFuel fitness metric), re-leased in early 2012. Far from being static signs, these traces are spun as highly evocative, active, and agentive concepts, and the complicated relay of technological projection becomes ever more difficult to unravel.

The analog and the digital, the human and the machine, the physiological and the emotive, the physical and the virtual are inextricably comingled; the terms seem to pollute one another in commercial discourse, and all we are left with is what Galloway (2006) humorously referred to as an "intense mi-metic thicket" (p. 316). Ideology thus becomes a free-floating signifier, a fantastical projection readily shaped by commercial imperatives, just as the critical terms seem to become meaningless. In the case of NikeFuel, event-based physiological signposts—the individual's rate of oxygen consumption and motion in a unique and rich spatio-temporal context—are captured as flattened industrial signs, ripped from their natural environment, and valued as branded measures of exchange. Practice as an activity in its own right gradually slips from our attention, and becomes merely a source of data to be

explained; to extend de Certeau's (1984) analysis of the activity of the fla-
neur, the poetic geography of the walker (or Nike runner) is undone by the
networked graph, as everyday understanding of experienced terrain yields to
a classificatory operation, and is successfully captured by discourse. The
SenseCam project, though a more positivistic pursuit, is an implicit parallel:
memory can be personally and socially transformative, but the clinical endeavor has an unarguably restorative function—the empowered subject is a
functional one.

The Networked Body

These various industrial commodities and research trajectories are signs of
the general renegotiation of autobiography in the digital age. Of course,
contemporary practices are neither simply birthed by industry, nor inherently
driven by technology; they are negotiated in the cultural field. As
individuals, we find refuge in biography, in the subjectivization and
individualization of the world around us. As willfully hypermediated
subjects we have become sophisticated archivists, and are now adept at
playing with the processes of signification that might situate and
individualize us. Socialization has been altered dramatically since the
introduction of networked communication, and, with it, traditional notions of
family and locality no longer seem to dominate the formation of our private
lives. No longer bound by these conventions, our personal attachments and
subjective longings have become individualized, freed from the moorings of
such outmoded contextual and placebound constraints. For many of us,
technophobia has given way to technophilia, even though common
generational distinctions still seem to inform the relative appeal of new
technologies. Robins (1995) suggested: "Rather than privileging 'new'
against 'old' images, we might think about them all—all those that are still
active, at least—in their contemporaneity" (p. 45). As we are presented with
new imaging technologies, we are also offered new ways of organizing the
visual field. But such new patterns, structures, and forms of organization
may find opposition in the cultures and traditions (the social contexts) that
ground them. The digital age, as with any significant evolutionary period,
necessitates looking forward and backward; it warrants a dual attention, an
understanding of both continuities and discontinuities.

Visuality is a trope of the networked gaze, yet it is still a powerful concept. The tension is invoked yet overlooked in a culture affected so broadly
by computerization. Chun (2004) noted: "The computer—that most non-

visual and nontransparent device—has paradoxically fostered 'visual culture' and 'transparency'" (p. 27). We must critically examine the limitations of transcoding within terminal space and understand what is revealed and what is concealed. The suggestion that networked communication has a high degree of indexicality is at odds with the actual operations of computation. "For computers to become transparency machines, the fact that they compute—that they *generate* text and images rather than merely represent or reproduce what exists elsewhere—must be forgotten" (Chun, 2004, p. 27, emphasis in original). The computer is a textual medium, though not a visual one, based on a technological language, and as such the founding principle of networked communication (and by extension the Internet) is control rather than freedom, for the controlling power lies in the technical protocols that make network communication possible.

Deleuze (1992) extends Foucault's periodization of the sovereign societies of the classical era (characterized by centralized power) and the disciplinary societies of the modern era (characterized by the bureaucratic mechanisms of institutions) by suggesting we have entered an age defined by societies of control: "Types of machines are easily matched with each type of society—not that machines are determining, but because they express those social forms capable of generating them and using them" (p. 6). The simple machines of the old societies of sovereignty have been replaced by machines of a third type: computers. Commodity fetishism is firmly embedded in this chronology, for as Deleuze (1992) suggested, this technological evolution must be understood as a mutation of capitalism. Whereas 19th-century capitalism is a "capitalism of concentration," consolidating power (acquiring the resources of production and property) within familiar enclosures such as the factory, Deleuze called contemporary capitalism "essentially dispersive": "The family, the school, the army, the factory are no longer the distinct analogical spaces that converge towards an owner—state or private power—but coded figures—deformable and transformable—of a single corporation that now has only stockholders" (1992, p. 6).

But we must not be overly reductive about the mechanics of power. To understand networked space as a space of either total control or total freedom is to lose sight of the paranoid impulse that frames our assessment, which is not directed at the technology but at the very operations of power. The paranoia stems from the reduction of political and social problems into technological ones (the desire for clarity), not only inducing paranoia, but also blinding us to how these very technologies operate and fail to operate, and forcing us into old ways of seeing—visuality returns with a vengeance as a safer form of investment, birthing willful acts of scopophilia. The networked gaze takes on a taxonomic function, especially as it attempts to transcode data into affect, an operation that only works if we (however incorrectly) place our unwavering faith in the image and the corresponding techniques of digital image capture.

Aligning image and affect, Behavior Image Capture is a therapeutic practice using digital image recording and wireless data transfer to record, store, annotate and review behavioral incidents. The technology has been used to observe children with autism and adults with PTSD, and captures behaviors in the patient's natural environment; these events can be shared with clinicians linked in a virtual treatment network. BI Capture was developed at the Georgia Institute of Technology, which began conducting research studies with an early prototype in 2006. The technology has since been licensed to Behavior Imaging Solutions; the company's suite of technology driven tools facilitates the observational, analytical, and collaborative needs of behavioral healthcare and special education professionals. The company's clients include the Department of Defense, State Departments of Education, behavioral health service providers, and several leading universities (Behavior Imaging Solutions, n.d.). The system's two components are Behavior Capture and Behavior Connect. Behavior Capture is an integrated behavior imaging tool including a high-definition Webcam, proprietary system software, and a wireless remote control. It captures and stores behavioral health events on video, and documents relevant events surrounding the behaviors. Behavior images are shared with remote professionals via Behavior Connect, an online portal enabling health and education professionals to interact with patients, specialists and other staff members, while building a library of shareable assets and a continuous health record. The system allows end users to capture behavioral incidents, record therapy and testing sessions for supervisory review, record custom videos for patients and caregivers, and annotate and organize a patient's corresponding video library.

Behavior imaging allows the capture, modeling, and analysis of social and communicative behaviors between children and their caregivers and peers. As a visual tool, these video-based analytics extend imaging technologies beyond the orthopedic base of X-Ray, and the neurological base of MRI and CT. The therapeutic feedback loop includes taking and annotating the video record, as well as developing computational methods and statistical models to analyze the visual, acoustic, and wearable sensor data. The goal is to develop new capabilities for the large-scale collection and interpretation of behavioral data, overcoming the statistical modeling challenges common to this field. To complement the video-based study of social interactions and parallel the work of trained psychologists, researchers aim to develop algorithms that can learn from data and make similar kinds of expert assessments, and automate the review of footage for detailed nuanced social interactions. The analytics are built on the acquisition of data about gaze behavior, facial expression, gestures, and audio modalities.

VizKid, a second behavior-imaging project led by researchers at Georgia Tech, is an information visualization system that supports the analysis of social orienting between two people interacting in a planned observation space. It has been used to study the avoidance behaviors of children with autism; following the strategic aim of Behavior Image Capture, the goal is to extract meaningful features from active behaviors recorded as image sequences. The system depends on an assessment room equipped with peripheral and overhead cameras that can map the world coordinates of the participating adult and child, reducing the subjects to a set of fiduciary markers and producing a historical overview of distance and mutual orientation by aggregating and visualizing the data from the more stable adult-centric perspective (Shin, Choi, Rozga, & Romero, 2011). The system has a therapeutic aim, but what does it mean to translate the body and subjectivity into an algorithm, and to align the clinical gaze with the industrial gaze of engineering, and the computational imperatives of computer science?

Algorithmic Culture

The AI research community has influenced the study and treatment of behavioral disorders by applying a computational perspective to human behavior, translating social interaction into data. From a scientific vantage point, the basic challenges are common sensing problems that have been extended to the translation of physiological variables, but the general theory is that all of these data modalities can be measured with the appropriate algorithms to process video and audio and wearable sensor streams, and that these modalities can be integrated to produce a composite portrait of the subject, moving from raw measurements of facial expression and voice to a statement about the child's emotional state (derived by measuring levels of emotion and engagement in task-oriented scenarios).

These queries, these looks, suggest a new industrial endeavor that extends the power/knowledge nexus embedded in earlier institutional practices. The desire to know in the context of the networked therapeutic relation is manifested as an attempt to ground disembodied subjects; there is a strong push toward embodiment, a process that in modern society has been achieved through a range of techniques including visual classification (photographic portraiture) and other tactics such as genetic mapping (which still has a visual register) (Lury, 1998). However, as Lury pointed out: "Having a (recognisable) body has historically not been sufficient to define an individual. Continuity of *consciousness* and *memory* are also necessary for a person to claim separate status as an individual" (p. 7, emphasis in original). Consciousness and memory are not produced by an accumulation of signifiers; rather, they are the products of narrative. As we approach the diagnostic trace, we may add up the formal cues and the system's specific forms of data, but we do so in order to construct a different type of portrait. Beyond the literal image, we create a projection by narrativizing the subject, or pulling the subject out of its immediate context and repurposing it. Differentiation, drawing distinctions between subjects, is accomplished only by indifferentiation, by taking their assets and subjecting them to an algorithm. The aspects of self that are socially determined are mapped alongside aspects of self that are biologically determined, and these variables are leveled, taken out of context, and fed into the database. Despite this apparent neutralization of what are fundamentally human properties as they take on the semblance of mathematical terms, one of the guiding principles of database construction is that relations should not be static but should instead abstract some portion of the real world that may itself change with time; the challenge is to hold onto the premise of dynamism. This prospect is lost when the subject is truly objectified—even subjective markers become objective factors. Individuals are rewritten as units of analysis that site designers can manipulate and users can analyze (Lury, 1998).

The development of hypertext and hypermedia and the translation of the body into a networked document depend on a foundational understanding of how language, speech, articulation, writing, and bodily copresence can be systematically remade by channeling them through a range of technologies subjected to the mechanics of engineering and the disciplinary logics of computerization. The networked identity is complex. Its composition may be planned, but the performance itself is always grounded in a specific context and connected to the protocols of networked communication. Lessig (1999) has made clear that code is a part of the regulatory process, although it is differentially applied; he suggests that different Internet architectures make behavior more—or less—regulable. We often forget the evolutionary nature of representational practice; focusing too much on ontological considerations, we fail to situate the medium—be it the videographic trace or the information network. The clinical studies of bodies and networks discussed herein are neither unique nor new. In 1948 Bell Labs began to explore fundamental problems of communication across circuitry—the discord between the semantic aspects of communication and the problems of engineering. "A Mathematical Theory of Communication," published by Claude Shannon in the 1948 issue of *The Bell System Technical Journal*, reads potential communiqués in statistical and logarithmic terms—transposing communication into numerical probabilities and signal-to-noise ratios.

Freeing up computation and freeing up culture requires a cooperative interaction that makes sense of the complexity of everyday life without necessitating reductionism or inflexible dependence on predetermined courses of action, understandings, laws of engagement, or programs. Code and data, and software and hardware, interact as a discursive construct (and each is a discursive construction in its own right) shaped by history, culture, industry, and technology that may encourage certain forms of engagement. But the outcomes of their applied uses may provide some useful lessons in understanding to what degree individuality is compromised by the general computational codes and filters of industrially sanctioned technological forms, or to what degree these codes can be massaged. Siri is at once localized and personalized, but she is also dispersed across more generalized and commercially predicated networked space; Siri *speaks* the network, the server, and the Web as she articulates a brand identity. As we relegate agency to a generating engine, framing our wants and needs in terms that Siri can understand (and learning to adapt in moments of failure), we must not lose sight of the dimensional itineraries that are the conditions of her possibility.

Notes

1. The concept of the subject is complex in critical theory, and there is a general tendency to complicate the notion of the autonomous individual capable of self-knowledge by introducing the shaping influences of discourse and ideology; such theories commonly suggest that the subject is best understood as a compromised position formed by the outward push of an innate root persona and the inward push of more universal cultural forces. Here, I suggest that technology must be added to the mix of decidedly exterior, cultural determinations that shape the subject.

2. The concept of postindustrial logic can be counterposed with the industrial logic that precedes it. Industrial logic (associated with the rise of the industrial age) aptly describes a factory or assembly line mode of production and offers waiting consumers a rather homogeneous product. Postindustrial logic suggests production on demand, inviting consumers into the production process by promoting either customization or postassembly adaptation.

3. This ad exploits an all-too-common narrative shorthand to suggest Apple's integrative yet globalizing push—the "camera" converses with Mac in her native Japanese, and he in turn speaks a universal hardware language—and imply a progressively commodified interdependence between western and eastern innovations in information technology.

References

Apple. (n.d.). *Siri. Your wish is its command.* Retrieved April 4, 2012, from http://www.apple.com/iphone/features/siri.html

Barthes, R. (1981). *Camera lucida: Reflections on photography* (R. Howard, Trans.). New York, NY: Noonday Press.

Behavior Imaging Solutions. (n.d.). *Company profile.* Retrieved March 3, 2012, from https://www.behaviorimaging.com/html/companyprofile.htm

Berry, E., Kapur, N., Williams, L., Hodges, S., Watson, P., Smyth, G., … Wood, K. (2007). The use of a wearable camera, SenseCam, as a pictorial diary to improve autobiographical memory in a patient with limbic encephalitis: A preliminary report. *Neuropsychological Rehabilitation, 17*(4–5), 582–601.

Bourdieu, P. (1990). *Photography: A middle-brow art* (S. Whiteside, Trans.). Stanford, CA: Stanford University Press.

Chun, W. H. K. (2004, Winter). On software, or the persistence of visual knowledge. *Grey Room, 18*, 26–51.

de Certeau, M. (1984). *The practice of everyday life* (S. F. Rendall, Trans.). Berkeley, CA: University of California Press.

Deleuze, G. (1992, Winter). Postscript on the societies of control. *October, 59*, 3–7.

Deleuze, G., & Guattari, F. (1987). *A thousand plateaus: Capitalism and schizophrenia* (B. Massumi, Trans.). Minneapolis, MN: University of Minnesota Press.

Englebart, D. C., & English, W. K. (1968, December 9). *A research center for augmenting human intellect* [Video file]. Menlo Park, CA: Stanford Research Institute. Retrieved April 30, 2012, from http://sloan.stanford.edu/MouseSite/1868Demo.html#complete

Foucault, M. (1970). *The order of things: An archaeology of the human sciences.* New York, NY: Vintage Books.

Freedman, E. (2011). *Transient images: Personal media in public frameworks.* Philadelphia, PA: Temple University Press.

Galloway, A. R. (2006). Language wants to be overlooked: On software and ideology. *Journal of Visual Culture, 5*(3), 315–331.

Greene, J. (2008, November 6). How Nike's social network sells to runners. *BusinessWeek.* Retrieved March 4, 2012, from http://www.businessweek.com/magazine/content/08_46/b4108074443945.htm

Lessig, L. (1999). *Code and other laws of cyberspace.* New York, NY: Basic

Books.

Lury, C. (1998). *Prosthetic culture: Photography, memory and identity*. New York, NY: Routledge.

Marcuse, H. (1964). *One-dimensional man: Studies in the ideology of advanced industrial society*. Boston, MA: Beacon Press.

McGuinness, D. L., Glass, A., Wolverton, M., & Pinheiro da Silva, P. (2007). Explaining task processing in cognitive assistants that learn. In *Technical report SS-07-04: Proceedings of the AAAI 2007 spring symposium on interaction challenges for intelligent assistants* (pp. 80–87). Menlo Park, CA: AAAI Press.

Panzarino, M. (2011, October 27). *Apple now has $81.5B in cash, 13.2M sq. ft. of facilities and 60K employees*. Retrieved February 28, 2012, from http://thenextweb.com/apple/2011/10/27/apple-now-has-81-5b-in-cash-13-2m-sq-ft-of-facilities-and-60k-employees

Rao, A. S., & Georgeff, M. P. (1991). Modeling rational agents within a BDI-architecture. In J. F. Allen, R. Fikes, & E. Sandewall (Eds.), *Proceedings of the second international conference on principles of knowledge representation and reasoning (KR'91)* (pp. 473–484). San Mateo, CA: Morgan Kaufmann.

Robins, K. (1995). Will image move us still? In M. Lister (Ed.), *The photographic image in digital culture* (pp. 29–50). New York, NY: Routledge.

Shannon, C. E. (1948, July, October). A mathematical theory of communication. *The Bell System Technical Journal, 27*, 379–423, 623–656.

Shin, G., Choi, T., Rozga, A., & Romero, M. (2011). VizKid: A behavior capture and visualization system of adult-child interaction. In G. Salvendy & M. J. Smith (Eds.), *Human interface and the management of information. Interacting with information. Symposium on human interface 2011 (HCI International 2011), Proceedings, Part II*. Lecture notes in computer science, Vol. 6772 (pp. 190–198). Berlin, Germany: Springer-Verlag.

Sontag, S. (1990). *On photography*. New York, NY: Anchor Books.

Weiser, M. (1991, September). The computer for the 21st century. *Scientific American, 265*(3), 94–104.

CHAPTER FIVE

Quest of the Magi: Playful Ideology and Demediation in *MagiQuest*

Paul Booth

> The standard path of the mythological adventure of the hero is a magnification of the formula represented in the rites of passage: separation—initiation—return: which might be named the nuclear unit of the monomyth. (Campbell, 1949/1973, p. 23)

The interactive role-playing adventure game *MagiQuest* offers an opportunity to explore the intersection of audience and production in a digital age, and at the same time, opens up new avenues of theoretical analysis within a digital media environment. *MagiQuest* is a live-action *Harry Potter*, a consumer-driven media spectacle that exists in an über-mediated environment of computers, digital graphics, on-screen animations, and interactive production. Players embark on quests to earn mystic runes, each rune a step on a larger path towards Magi-hood. The eventual attainment of the rank *Master Magi* comes through perseverance, attention to detail, and a strong sense of moral fortitude. Each Master Magi possesses a stout heart, an honorable moral code, and a knowledge of magic unrivaled throughout the kingdom.

To become a Magi, one must learn to *think* like a Magi. In this way, *MagiQuest* is a pedagogical game. The quests presented by *MagiQuest* challenge and thus teach participants; the lessons learned are often more ideological in nature than they are practical. In this chapter, I reflect on how *MagiQuest* functions both as a game and as an immersive environment, explore how *MagiQuest* can be analyzed using a *demediated* look at the game environment, and demonstrate the rough dialogue between audience and production *MagiQuest* navigates. I argue that *MagiQuest* integrates a productive user experience into a pedagogically enticing environment. The *MagiQuest* environment provides a glimpse into a world of ubiquitous mediation, and the über-mediation of the game teaches the player how to function within our mediated world.

But let's use a Time Rune to reverse the path of this chapter a bit and see where we started. To non-Magi, *MagiQuest* is a game conceived by child psychologist Denise Weston and played in an immersive environment by those enamored of the classic Campbellian Hero myth. Players create their own hero character, attach the character to a clan, and purchase a magic wand (an electronic infrared, or IR, device) that they can use to activate particular aspects of the gameplay through a unique identifying code tied to their character. On top of each wand sits a topper, a purchasable customization that can change players' abilities. Abilities are earned by completing quests and gaining runes: mystic skills for the players' characters to manifest. Sample runes include the Reveal Rune, which allows characters to "see what would be invisible," or the Ice Arrow Rune, which can "penetrate the thick red scales of the Red Dragon."[1] Runes also allow players to duel, or battle another player in a videogame-like environment.

To earn runes, players embark on quests, which always begin in front of a computer monitor. Waving the wand in front of the monitor activates a sensor that reads the wand's signal. The computer recognizes the player-character's unique code, and offers a range of quests to play. Each quest sends players searching for particular markers or landmarks in the *MagiQuest* space in order to earn runes. Players record their finds by waving their wands towards the markers. The wand picks up the signal, records the player's status, and downloads that personal information to a database. Upon finding all markers for a particular quest, the player returns to a central computer to collect the rune and begin another quest.

A personal example may elucidate this process. To earn the Lightning Rune, or the ability to use an attack spell (one of the first spells one should learn in *MagiQuest*), I stood (in the guise of my character, Sir Paul the Wise of the Trixter Clan) at the central core computer area and waved my wand topped with the Master Magi Topper ("The Master Magi Topper holds the power of the Learned Owl"). An animated Learned Owl appeared on the screen, flapped its wings, and then dissipated to reveal a number of options. I touched "Quest" and then selected the one I wanted—Lightning Rune. A live-action wizard appeared on the screen and introduced my quest by speaking a verse that posed a simple riddle: e.g., "If you be the type/to seek out a **good book**/near the quest stones, by the entry/I'd urge you to look." I turned around and saw a large tome near me. I shook my wand, a *woosh* sound was emitted, and the book lit up with sparkling lights. I had found the first clue. Three more clues pointed to a sword in the stone, a shield, and a suit of armor, each of which lit up when I shook my wand near them. Upon finding the last clue, I returned to my initial computer, waved my wand, and the wizard appeared to congratulate me and offer me the Lightning Rune (I later used the Lightning Rune to challenge my wife to a duel, which I lost). I touched the screen to accept the rune and began my next quest (the Enchant Creature Rune). Each subsequent quest asks new riddles, some of which are quite challenging—one quest (Master Magi Rune Quest) took me the better part of 45 minutes. My location (Lombard, Illinois) offers 12 quests and three longer, more complex adventures; different locations offer alternative quests, different runes, and separate adventures.

Key to any adventure or quest in *MagiQuest* is the immersion of the participants; it is with this in mind that I begin to discuss some of the theoretical consequences of the *MagiQuest* experience. *MagiQuest*, described online as "an interactive live-action, role playing game," is played by individuals in an immersive environment. In the location closest to me, for example, the *MagiQuest* environment is housed in the lower level of a mall, and takes up the space of roughly four stores. Split into nine areas—it stretches from the Twisted Woods near Pixie's Perch to the Dungeon at the Dragon's Lair—the room houses a number of stanchions and scaffolds to create a labyrinthine interior. To play a quest, one must traverse across the maze a number of times, creating not just a simulation of a hero's quest, but also a remarkably good workout. Not all *MagiQuest* locations are the same; for example, the multiple locations housed in Great Wolf Lodges take participants outside of the central *MagiQuest* zone to interact with objects in the hotel itself. Similarly, my location has quests that sent me out to find musical instruments scattered across the mall. The immersion of *MagiQuest*, therefore, goes beyond a closed stage for playing out medieval fantasies, spilling out into malls, hotels, and theme parks.

In a hyperreal Baudrillardian sense, then, the bizarre falseness of these not-so "real world" locations such as the mall, hotel and theme park lends what I have previously called a "demediation" to the *MagiQuest* world (Booth, 2010). In what follows I show how a consumer's immersion in the *MagiQuest* world leads to a greater *sense* of active production (whether or not that sense is accurate). *MagiQuest* uses the dichotomy between production and consumption to teach ideological lessons to the players. Questing becomes a metaphor for life, and naturalizes the concepts of commodification and conflict for the players.

MagiQuest and Demediation

The hero adventures out of the land we know into darkness; there he accomplishes his adventure, or again is simply lost to us, imprisoned, or in danger; and his return is described as a coming back out of that yonder zone. Nevertheless—and here is a great key to the understanding of myth and symbol—the two kingdoms are actually one. (Campbell, 1949/1973, p. 217)

MagiQuest links the mediated world with the non-mediated world, but in doing so it creates a *demediated* space where technology exists seamlessly with the audience production of the game. Demediation mirrors the way our lives today display multiple media as a whole environment. With demediation, the ubiquitous multitude of mediations in the digital environment envelops the user. Yet, importantly, the media don't disappear from the user's view, but remain still *knowable* as media. I have previously (Booth, 2010) defined demediation as an intermedial position between hypermediation and immediacy—two terms from Bolter and Grusin (2000). Hypermediacy is the state wherein media use is *obvious* and we look *at* the media—for example, we notice the special effects of a film instead of seeing them as part of the film's world. Immediacy, on the other hand, defines the logic of erasure, where we forget we're looking at media and see *through* it instead (Murray, 1997). The dichotomy between hypermediacy and immediacy comes to a head, for example, when watching a DVD on one's computer. Given the multiple levels of mediation in this experience, it's easy to see this viewing as *hyper*mediated: the DVD controls appear at the bottom of the screen, we can half-size the window and type notes when watching, or our email notification dings so we check Gmail while the movie is playing in the background. We are aware of the DVD *as media*. Yet, it's also surprisingly easy to sit back in the chair, resize to full screen, and become lost in the telling of the story: to forget that what we're watching is a movie, and to see through the screen into an immediate experience.

Bolter and Grusin (2000) dealt almost exclusively with singular in-stances of mediation—the television, the photograph, the computer, and so forth—as separate entities. But mediation in our digital environment is rarely experienced in a media vacuum. That is, our mediated environment contains multiple media technologies, surrounding us and creating vast arrays of many different types of hypermediation as well as many different levels of immersion all at once. We see through some, we look at others. Demedia-tion—where mediation is ubiquitous, obvious, and at the same time ef-faced—is a more fitting moniker for the experience of today's media, one represented and exemplified by *MagiQuest*.

I've previously used the example of Alternate Reality Games (ARGs) to exemplify demediation, and in many ways, *MagiQuest* is similar to these real-world, immersive, game-like experiences. An ARG is a game played in the real world, and is, as game researcher Jane McGonigal described, "anti-escapist" (2011, p.125). That is, ARGs use everyday technology and real-world interactions to engage reality with game-like results. Indeed, that sense of engagement complements what games researcher Yam San Chee (2007) defined as the three most important components of immersive gaming—embodiment, embeddedness, and experience—all of which engage players in "learning by doing, observing the outcomes of [their] actions, testing [their] hypotheses about the world, and reflecting further on [their] own understanding" (p. 15). Similarly, Schroeder (1996), citing Huizinga (1938/1950), argued that all play engages players and "operates within its own space" (p. 147). This space, called at times the "Magic Circle" (Huizinga, 1938/1950; Salen & Zimmerman, 2004), a "playspace" (Castronova, 2004; Schroeder, 1996), a "third space" (James et al., 2009) or the "game space" (Juul, 2005), exists as part of our real world, but with separate rules, activities, and actions. For example, when one enters the game world of *tag*, the rules state that one should slap opponents to tag them out of that round. In the real world, tagging someone runs the risk of assault and battery charges.

Like large-scale, Internet-based scavenger hunts, ARGs send players shuffling between the real world and the game world to find clues and solve puzzles (Sparks, 2009). ARGs rely on the distinction between real and non-real, and attempt to hide the Magic Circle of the game world within the environment of the real world. It's important to note that ARGs use everyday technology to straddle this divide between the real and the gameworld—and because the technology *is* everyday, it disappears through its very ubiquity.

> Despite the appearance that they blur the boundaries between the ludic and the non-ludic, ARGs actually *reinforce* that boundary; and in doing so, they…provide a reassurance and strengthening of what contemporary media studies might call the unmediated reality of physical existence. (Booth, 2010, p. 181, emphasis added)

Using everyday media to hide the game serves a dual function that reverses common assumptions about games: they both immerse the player in the game world and also make obvious the mechanics of the real world. In other words,

> the mediation of the ARG cannot be mediation because there is no "game" to mediate outside of the ARG. ARGs exist because they reverse and exceed immediacy and hypermediacy: they demediate both the non-ludic and the ludic. The ARG thus uses media to hide its mediation; as such, the ARG seems immersive, but effaces the mediation it necessarily presents. (Booth, 2010, p. 187)

Like an ARG, the magical kingdom of *MagiQuest* is composed almost entirely of elements of mediation, from the IR wands wielded by the Magi to the computer terminals scattered throughout the environment. Wave your wand at the wall and any number of exciting sound effects, images, lights, or movement may occur. Computer terminals are posted every few feet, waiting for a flick of the wand to activate and display whatever creature lives in them. For example, to complete the Freezing Rune quest, one must collect a number of items around the kingdom and then find the computer terminal in which the Old Man in the Stump lives. As the player approaches the computer, a monitor reveals images of the Old Man napping; when the wand is waved in front of the terminal, he awakens and demands to know who disturbed him. Terminals such as these are scattered around the environment, and even if it is the end of their quest, users can always shake a wand at a computer to get hints, see their current standing, or discover what else they have left to do. Even beyond the immediate spaces within *MagiQuest*, the game is replete with mediation. Computer monitors posted in the Market Place (the large area at the front of any *MagiQuest* kingdom, in which players can purchase costumes, wand accessories, games, models, and any manner of other accouterments for their game play) display current and all-time high standings of the players. There is even an online version of the game, *MQOnline*, which displays a player-designed avatar in the magical kingdom, facing various online quests, à la *World of Warcraft* (minus the fighting). Crucially, at least at the location nearest to me, there is a quest within the live *MagiQuest* game that requires *MQ Online* to function.

Demediation thus exists as an integral part of the mediated immersion of *MagiQuest*. However, unlike hyperreality, another theory of ubiquitous mediation, demediation invokes a practical and applicable theory within the game. For Baudrillard (1994), hyperreality exists as a state of culture completely enveloped in mediation, so much so that reality itself disappears. Mediation overtakes all else. All that's left is a sense of mediated truth with no basis in what Baudrillard called "first-order reality" (p. 6). There is no antecedent to the simulation; it is all that exists. So, a mediated event takes the place of the real event, and becomes more real than that event. We learn more about the experiences of war from war *films* than we do from actual combat. Importantly, one consequence of hyperreality is that the inherent falseness of the world is disguised by the fact that obvious hyperreal environments exist in microcosm. Baudrillard famously used the example of Disneyland to say that the Magic Kingdom exists to disguise the fact that the rest of the world is just as simulated as Disneyland is. We see that Disneyland is false (and enjoy it for its falseness), and then don't notice that everything else around us is equally false.

Schroeder (1996, p. 144), however, has admitted that "using Baudrillard (and Baudrillard-style theories of simulation) as a theoretical tool is a problem"—and one that demediation sidesteps. Indeed, we enter into this problematic interaction with the media environment when we examine *MagiQuest* from this Baudrillardian standpoint. Just as "the boundaries of the playspace begin to disappear," when games become immersive, so too does any theoretical analysis of that game, from a "hyperreal" standpoint (Schroeder, 1996, p. 148). *MagiQuest* is an obvious hyperreal event, but because it's played in hotels, malls, and theme parks (which can also be analyzed as hyperreal locations), the hyperreality becomes subsumed by more hyperreality. If we are in the age of simulation, as Cubitt (2001) noted, "we can no longer distinguish between play and work" (p. 133). Given this shift, a hyperreal analysis would lose all defining characteristics, revealing not the loss of the real, but rather the requisite loss of the hyperreal. This effectively effaces hyperreality through its own hyperreal analysis.

Instead of venturing down this path of theoretical tautology, I think it's more explanatory to focus on the interaction of the media technology interface and the user (Stein, 2006). *MagiQuest* is demediate in that the ubiquity of the mediation within the game teaches players how to use technology in an intuitive and instrumental way. The interface disappears through its obviousness, and the wand—much like the Nintendo Wii controller, the Xbox Kinect, or the PlayStation Move—becomes the McLuhanesque mediated extension of the human body. Media are internalized; reality is not sublimated. Players become one aspect of the mediated environment, a technology in and of themselves, focusing the gameplay away from a purely deterministic dichotomy between the interface and the player, and toward an environment in which the player becomes part of the mediation itself. The hypermediation of the game *becomes* an immediate experience through immersive gameplay. Ultimately, *MagiQuest* highlights a shift in media theory, because the mediation of the game stands separated not from other mediation, nor from the players, but from a demediated reality. MagicQuest becomes inextricably enmeshed with the non-mediated world as merely one component of an already convoluted co-created player activity.

The Interpellation of MagiQuest

With the personifications of his destiny to guide and aid him, the hero goes forward in his adventures until he comes to the 'threshold guardian' at the entrance to the zone of magnified power.... Beyond them is darkness, the unknown, and danger. (Campbell, 1949/1973, p. 77)

As a demediated text, then, *MagiQuest* highlights contemporary media theory. But what is the practical result of this demediation of *MagiQuest*? First and foremost, *MagiQuest* can perhaps best be described as a *game*— one in which participants become fully immersed as role-players (Grant, 2010; Hutchens, 2008; Salas, 2009). To pretend is key to playing *MagiQuest,* and players must role-play as characters in order for the game to be successful (Salas, 2009). Like improvised theater, *MagiQuest* creates a story as it is being played—players help construct the series of events as events are encountered (Uren, 2007). One can't play if one is self-conscious about shaking a plastic wand at a painted picture of slime on a plywood tree (see Kushner, 2009, as an example of someone who doesn't become immersed). We must be aware of the media to play, but we must ignore it to be immersed.

MagiQuest also illustrates important pedagogical ramifications of game-based learning. Through its demediation, *MagiQuest* interpellates key ideo-logical lessons within its game, encouraging *teamwork, imagination,* and *play.* According to a press release from its creator (Creative Kingdoms), *MagiQuest* was designed to "get kids off their couches...moving and interacting together." Through its interactive mechanisms—the IR wands, the touch screens, and the immersive environment—*MagiQuest* stimulates the same learning capabilities in children as do games (Schoolfield, 2006).

Indeed, the effectiveness of play as a pedagogical tool has been well documented. For example, Chee (2007) described how game-based learning helps students learn more information, retain that information longer, and apply that information in different settings. Of today's generation of students, he wrote, "school is only one source of learning; out-of-school learning and informal learning increasingly account for more of a millenial's overall education" (p. 5). Video-game researcher James Paul Gee (2003) argued that new times call for new literacies: the hypermediated environment of video games, cell phones, PDAs, iPads and TiVo demands a new type of learning, one predicated on new "semiotic domains" made up of "images, sounds, gestures, movements, graphs, diagrams, equations, objects, even people" (p. 17)—all aspects of *MagiQuest*. Focusing on play as an integrated method of pedagogical practice fosters these new literacies, forging new students who grow up knowing how to act, interact, and react in a 21st century media environment. Furthermore, James et al. (2009) have shown that many leisure activities in "the new media" online, including "gaming [and] instant messaging, social networking on Facebook and MySpace, participation in fan fiction groups, blogging, and content creation" can all be subsumed under the category "play," contributing to a new understanding of online activity (p. 11). Jenkins et al. (2009) found that this type of play, as a concept situated within teaching and learning practices, is key to "shaping children's relationships to their bodies, tools, communities, surroundings, and knowledge" (p. 35). Play is crucial to learning because "this activity is deeply motivated. The individual is willing to go through the grind because there is a goal or purpose that matters to the person. When that happens, individuals are engaged" (p. 37). In fact, using games as pedagogical models has been shown to increase retention in schools (Hernandez, 2009). Play allows children (of all ages) to learn in a safe environment. Failure in a playspace merely results in sitting out the next round, whereas failure in the real world can entail punitive action, castigation, or punishment. Playspaces allow players to be stimulated, to experiment, to be social. These informal, third spaces allow "conduct that is both meaningful and engaging to the participant and responsible to others" (James et al., 2009, p. 15).

Playspaces are effective not only because they offer safe zones, however, but also because they are put into practice through players' interactive efforts. Playspaces are the ultimate demediation: they exist as created by the people who participate in them, and are both obvious and immersive. Thus playspaces, and the games residing within them, exemplify the confluence of demediation that commands attention in today's convergence culture, where users both produce and consume media on the same device, and "participation is understood as part of the normal ways that media operate" (Jenkins, 2006a, p. 246). *MagiQuest* relies on this co-creation: Creative Kingdoms produces an environment and interactive media; players of *MagiQuest* produce their own adventures through role-play, interaction, and imagination within the environmental realm (Schoolfield, 2006). Indeed, important nonscholastic skills are also built, as Adrian (2007) noted of *MagiQuest* that "*recognition*, a much sought after societal quest in itself, is instantly accomplished" (p. 46, emphasis added). That Creative Kingdoms makes money from players' own creativity invites additional critique, as part of today's "digi-gratis" economy (Booth, 2010).

Playspaces also offer a unique opportunity for safe and effective pedagogical practices. I have previously stated that we have entered an era defined by a particular "philosophy of playfulness," where "the contemporary media scene is complex, and rapidly becoming dependent on a culture of ludism" (Booth, 2010, p. 2). As an exemplar of how play is becoming a ubiquitous marker of this cultural ludism, *MagiQuest*, I argue, enacts a further refinement in the nature of these new literacies, and provides a deepening understanding of changes to the media environment. Whereas a video game literacy, as described by Gee (2003), Chee (2007), or Nielsen, Smith, and Tosca (2010), might provide a solid background on new movements in media studies, the type of literacy engendered by *MagiQuest* offers a more encompassing venture into a larger digital environment.

Specifically, by straddling the audience/producer chasm, *MagiQuest* embodies a playful learning environment that teaches children not only to be active *producers* of information and play, but also to become passive *consumers* of media information. As Jenkins (2006b) has warned us, as freeing as new technology may seem for democratic and consumer-driven media (and, indeed, *MagiQuest* seems the ultimate in liberating technology), these technologies are still constructed and run by large corporations. To be fully engaged in the material, one must embody the true Magi, a role Creative Kingdoms is only too happy to illustrate and present as the norm. That is, *MagiQuest*'s veneer of pedagogy actually ends up teaching two different types of activities. On the surface, the player's active, imaginative engagement with the material does provide a safe space for creative immersion; underneath that immersion, however, the type of play engendered by *MagiQuest* interpellates particular norms of behavior.

The Althusserian notion of interpellation describes the process wherein an identity is imposed by an external force onto a subject—one thus becomes "constitutive of all ideology" of that force (2005, p. 84). That is, the subject in an ideological state is "hailed," or prescribed a specific set of characteristics by this external force. Althusser's classic example of interpellation—the police officer yelling "Hey, you there!" and everyone turning around, thinking she's talking to *them*—still has resonance today. As Greg Smith (2010) has noted, "You accept an identity not of your own making but one that the officer offers: You are a 'possible suspect'" (p. 90). The media hail us from wherever we are to point us toward what our identities, our subjectivities, seem to be. From advertisements to television programs, from Hollywood films to public radio, American media interpellate what it means to be (to act like) an American, and, as Stuart Hall (1996) said, "place [us] as the social subjects of particular discourse" (p. 5). "Bit by bit," claimed Smith (2010), "this positions us in society as members of *this* group but not *that* one" (p. 91, emphasis added).

MagiQuest offers a relatively easy vantage point from which to observe some of the interpellation that occurs in the game. From the obvious gender norms that are displayed with the characters—the Princess is traditionally beautiful, tall, blonde, and thin, and the Lady in the Leaves (the keeper of the Dazzle Rune) is highly sexualized, especially considering the youth of the average player—to the lack of any ethnic or racial diversity among the on-screen actors, *MagiQuest* seems to standardize the magical kingdom for beautiful, white people. Participants are told that they are Magi, true and virtuous adventurers in a magical land. But looking around that land, players are hailed as Magi that fit a particular vision, a particular image, and a particular act. The sneaky part of this hailing occurs when participants aren't just told that this is how to look in the kingdom, but rather actively co-create this image, both by immersing themselves into the demediate world of *MagiQuest* and by purchasing *MagiQuest* merchandise. The interactivity of the game helps foster the implied messages of the game (Brookey & Booth, 2006). Players both implicitly and explicitly buy into the ideological standardization portrayed through the game.

And, in fact, this buying into *MagiQuest* leads to a more complex understanding of the interpellation within the game. Beyond the superficial aspects of the game, players also encounter a general metaphor for ways to live their lives. Players' engagement with the immersive aspects of the game, through their own co-creation of their game experience, helps drive their imbibing of the game's ulterior mythology. As Rosenbloom (2003) described it, this very immersion helps solidify the teachings of any game in the minds of the players: "the driving force in these immersive, realistic environments is the user's experience" (p. 31). *MagiQuest* uses the game environment to normalize and facilitate a standard narrative that players take as given in a world created as fiction. It's not hard to see the truth of this fictionality, especially when that fiction is presented in a participant-focused environment. Jesper Juul (2005), for instance, has shown how fictional worlds can help shape and influence a player's understanding of the real world. In his *Half-Real*, he showed how "the fictional world cues the players into making assumptions about the real world in which the player plays the game" (p. 168). So, for example, whereas we all know that there aren't wizards, magical owls, and dragons in the real world, the fact that they are real in the *MagiQuest* game as we play it gives them an air of authenticity, which means that their lessons are taken more to the heart than if we were simply told about them outside that environment. Playing becomes doing becomes being.

But what are these lessons, these hidden stories that *MagiQuest* instills in the minds of its players? Perhaps the most potent message *MagiQuest* offers is that consumption and competition are normal, if not crucial, aspects of modern-day cultural practices. Players are hailed as Magi before they even enter the kingdom, and are then told that Magi fit into specific, heroic stereotypes that can be commoditized. For example, when a Magi enters the kingdom, the first thing he/she does is fill out a card with his/her moniker, vital statistics, and choice of clan: Majestic, Shadow, Trixter, Warrior, or Woodsy. As shown on the *MagiQuest* Web site, each clan is represented by its own wand topper, with special powers granted only to those who spend 15 dollars to attach it to their wand. Magi are never told that they *must* purchase the topper; merely that the topper is a chance to "express [their] game character" and "personalize [their] MagiQuest wand." Other toppers invoke different abilities, and are available in the Market Place.

I certainly don't wish to make it sound as though this is a negative or even an unwelcome aspect of the game; speaking personally, having different wand toppers varies and gives a certain flair to the gameplay (I particularly like my Trixter wand topper, which looks like a gyroscope, lights up orange when flicked, and makes a magical noise). Indeed, great amusement was had by all my companions one day when I purchased the "wand extender," a topper that exists solely as an exemplar Freudian symbol. But the *normalization* of the process, the fact that consumption is a matter-of-fact aspect of the game, and is co-created by the players just as much as it is offered by Creative Kingdoms (no one *has* to purchase a topper, but I have yet to see someone who hasn't), makes this an inarguably central metaphor of the game. Indeed, *MagiQuest* was originally conceived to be centered on commodification; creator Weston stated that "There are multiple sources of revenue coming in.... 'We get you in the very beginning' when you buy the wand" (as cited in Fleisher, 2007).

The consumption of *MagiQuest* extends into the game world as well, supporting Juul's (2005) point that game experiences mirror real-world ones. Within the game, players accumulate two things: Experience Points (XP), the accrual of which leads to promotion from mere Magi to Master Magi, and gold. Gold is earned by flicking one's wand towards magic chests of money, which open, sparkle, and offer their contents to the player. "You've found 200 gold!" they exclaim. The increase in wealth is recorded via the player's wand and is factored into the scores at the end of the game: the more wealth (and XP) one accrues, the higher one's social standing—a clear metaphor for climbing the socio-economic ladder.

It is impossible to transfer gold from one person to another, so sharing is hardly an option. Indeed, the initial separation into clans represents another social interpellation on the part of *MagiQuest*—the clear differentiation and segmentation into social groups. Although no one group is seen more positively than another, the distinctions between groups highlight a deterministic view of social relations (mirrored in *MagiQuest*'s antecedent, *Harry Potter*, in which the Hogwarts' matriculates are sorted into four houses, each with clear-cut characteristics). Differences are connoted by the language describing each clan, and rarely affect actual gameplay; they simply help structure and identify the players. The Woodsy clan, for example, is peopled with "creatures who can sometimes be a little mischievous." Trixters are "wise-cracking jokers and jesters who use music as part of their practical joking." Members of the Shadow clan "tend to be sneaky Magi who are usually up to no good." As Gee (2003) described, the three identities we create when we play games—the virtual identity of the character, the real identity of the player, and the "projective identity" that describes the "interface be-tween…the real-world person and the virtual character"—all interact as a "larger whole" (pp. 55–56). *MagiQuest* players can envision themselves in this projective identity through contrasting the same characteristic against others with that same trait—the negative connotation of the Shadow clan's "sneaky" mentality is similar to the more positively weighted "mischievous" tendency of the Woodsy clan. This articulates one particular virtual identity, affecting the "real-world" identity of the player. Identity formation in games is crucial (Chee, 2007), especially for children. Narrowly defining these characteristics creates associations that could have repercussions in the real world as well.

Indeed, what the clan segmentation and the game commodification have in common is the fostering of a sense of competition in the game. Although players are not deliberately pitted against each other (except in the separate dueling part of the game), they are often implicitly set in contrast, from the competing clans to the accumulation of XP and gold. The only cooperation truly experienced in the game is at the end, when more than one person can take on the final boss, the dragon. Even that level of teamwork is optional.

In this way, the quest of MagiQuest becomes a metaphor for the trap-pings of a constituted, capitalistic life. Players engender the characters within the game, and are led through a series of quests that inexhaustibly lead to ac-tive consumption and implicit competition. Yet, players get in on the act too, becoming so immersed in the game that the interpellative forces at work be-come naturalized. But perhaps the most important ideological lesson im-parted by *MagiQuest* is the influence of the pervasiveness and ubiquity of the contemporary media environment on the players themselves. Articles and re-ports about *MagiQuest* continually highlight the virtual reality aspects of the game (Salas, 2009; Szadkowski, 2007) and its similarity to video games (Kushner, 2009; Salas, 2009). *MagiQuest* inherently relies on digital tech-nology to foster the immersion of the players in the game and then attempts to hide this mediation from the participants through intuitive and seamless interfaces. The demediation of the game highlights how *MagiQuest* teaches us how to live/work/play/live/be a mediated life.

The End of the Quest

> When the Hero-Quest has been accomplished, through penetration to the source, or through the grace of some male or female, human or animal personification, the adventurer still must return with his life-transmuting trophy. (Campbell, 1949/1973, p. 167)

MagiQuest offers players a number of things: quests and adventures, wands and magic, interaction and immersion. But *MagiQuest* also gives its players certain rules and lessons by which it instructs everyday interactions. What happens when questing is taken as a metaphor for life? What happens when overt mediation becomes ubiquitous? What happens, then, when the game ends and Magi move on to other aspects of their life? As Chee (2007) has noted, any discussion of games must include a discussion of the "set of values and beliefs" that "are manifested through what the game seeks to convey via its design." In other words, "games are often a rhetorical medium" (p. 26). If that is the case, then *MagiQuest* tells us about life in a post-industrial, digital society where mediation is central to contemporary existence.

All quests end; the classic hero's quest could end in victory, redemption, or death. But when *MagiQuest* ends, players take away important metaphors that help structure and explain today's mediated world. That players become co-creators of this world through their gameplay fosters a greater persuasive function within the game mechanics. As the gameplay becomes normalized, as players scribe their own meaning into the game, it becomes more immersive and influential. This is not true of all contemporary games, most of which exist as mere products to purchase (Kerr, 2006). Games are a singular entity existing as a purchasable piece of media—the Bolter and Grusin-esque (2000) epitome of remediation. *MagiQuest*, although often described using video game terminology, represents a more complex mediated experience. How best, then, to describe a game that is more a *practice*, and less a *product*? Rather than focus on how *MagiQuest* harnesses virtual reality, simulation, and hyperreality to set it apart from the "real-world," it is more instructive, and more accurate, to show the ways *MagiQuest* demediates, invoking characteristics that highlight its connection to our everyday life. *MagiQuest* may be a representation of a magical, mythological kingdom, replete with princesses, goblins, dragons, and pixies. But in many ways, it is also an extension and remarkable facsimile of our own digital world.

Note

1. Unless otherwise stated, all quotations about the MagiQuest game mechanics and shop come from the *MagiQuest* Web site, www.magiquest.com.

References

Adrian, C. (2007). Families fall under the spell. *Tourist Attractions and Parks, 37*(5), 44–46.

Althusser, L. (2005). Ideology and ideological state apparatuses (Notes towards an investigation) (B. Brewster, Trans.). In M. G. Durham & D. M. Kellner (Eds.), *Media and cultural studies: Keyworks* (Rev. ed.) (pp. 79–88). Malden, MA: Wiley-Blackwell. (Original work published 1970).

Baudrillard, J. (1994). *Simulacra and simulation* (S. F. Glaser, Trans.). Ann Arbor, MI: University of Michigan Press. (Original work published 1981).

Bolter, J. D., & Grusin, R. (2000). *Remediation: Understanding new media.* Cambridge, MA: MIT Press.

Booth, P. (2010). *Digital fandom: New media studies.* New York, NY: Peter Lang.

Brookey, R., & Booth, P. (2006). Restricted play: Synergy and the limits of interactivity in *The Lord of the Rings: The Return of the King* videogame. *Games and Culture, 1*(3), 214–230.

Campbell, J. (1949/1973). *The hero with a thousand faces.* Princeton, NJ: Princeton University Press.

Castronova, E. (2004, August 3). *Virtual world economy: It's Namibia, basically.* Retrieved February 6, 2012, from http://terranova. blogs.com/terra_nova/2004/08/virtual_world_e.html

Chee, Y. S. (2007). Embodiment, embeddedness, and experience: Game-based learning and the construction of identity. *Research and Practice in Technology Enhanced Learning, 2*(1), 3–30.

Cubitt, S. (2001). *Simulation and social theory.* London, England: Sage.

Fleisher, L. (2007, November 14). MagiQuest's creator franchises MB-first game in U.S., Japan. *Sun News.* Retrieved February 26, 2011, from http://www.myrtlebeachonline.com/business/story/249721.html

Gee, J. P. (2003). *What video games have to teach us about learning and literacy.* New York, NY: Palgrave Macmillan.

Grant, T. (2010, December 21). At Great Wolf Lodge, some grown-up time. *The Washington Post,* T39. Retrieved February 26, 2011, from http://www.washingtonpost.com/wp-dyn/content/article/2010/12/21/AR2010122105597.html

Hall, S. (1996). Introduction: Who needs 'identity'? In S. Hall & P. du Gay (Eds.), *Questions of cultural identity* (pp. 1–17). London, England: Sage.

Hernandez, D. (2009, Dec). Gaming+autonomy=academic achievement. *Principal Leadership, 10*(4), 44–48.

Huizinga, J. (1950). *Homo ludens: A study of the play element in culture.* Boston, MA: Beacon Press. (Original work published in 1938).

Hutchens, B. (2008, July 21). Great Wolf Lodge resort much more than just a water park. *The News Tribune.* Retrieved February 6, 2012, from http://www.tmcnet.com/usubmit/2008/07/21/3557900.htm

James, C., with Davis, K., Flores, A., Francis, J. M., Pettingill, L., Rundle, M., & Gardner, H. (2009). *Young people, ethics, and the new digital media: A synthesis from the GoodPlay Project.* Cambridge, MA: MIT Press.

Jenkins, H. (2006a). *Convergence culture: Where old and new media collide.* New York, NY: New York University Press.

Jenkins, H. (2006b). Interactive audiences? The "collective intelligence" of media fans. In *Fans, bloggers, and gamers: Exploring participatory culture* (pp. 134–152). New York, NY: New York University Press.

Jenkins, H., with Purushotma, R., Weigel, M., Clinton, K., & Robison, A. J. (2009). *Confronting the challenges of participatory culture: Media education for the 21st century.* Cambridge, MA: MIT Press.

Juul, J. (2005). Half-real: Video games between real rules and fictional worlds. Cambridge, MA: MIT Press.

Kerr, A. (2006). *The business and culture of digital games: Gamework/gameplay.* Thousand Oaks, CA: Sage.

Kushner, D. (2009, September). Escape to Middle-Earth. *Wired, 17*(9), 72.

McGonigal, J. (2011). *Reality is broken: Why games make us better and how they can change the world.* New York, NY: Penguin.

Murray, J. (1997). *Hamlet on the holodeck: The future of narrative in cyberspace.* Cambridge, MA: MIT Press.

Nielsen, S. E., Smith, J. H., & Tosca, S. P. (2010). *Understanding video games: The essential introduction.* New York, NY: Routledge.

Rosenbloom, A. (2003). A game experience in every application. *Communications of the ACM, 46*(7), 28–31.

Salas, R. (2009, October 12). Enter the dragon: Grab a wand and hunt all manner of fantasy creatures at MagiQuest, a new interactive adventure at the Mall of America. *Star Tribune.* Retrieved February 6, 2012, from http://www.startribune.com/lifestyle/64041862.html?page=all&prepage=1&c=y#continue

Salen, K., & Zimmerman, E. (2004). *Rules of play: Game design fundamentals.* Cambridge, MA: MIT Press.

Schoolfield, J. (2006). Quest masters. *Funworld, 22*(10), 42–51.

Schroeder, R. (1996). Playspace invaders: Huizinga, Baudrillard and video game violence. *Journal of Popular Culture, 30*(3), 143–153.

Smith, G. (2010). *What media classes really want to discuss: A student guide.* New York, NY: Routledge.

Sparks, K. A. (2009). Will your next job be in an alternate reality? *Canadian Screenwriter, 11*(2), 27.

Stein, L. E. (2006). "This dratted thing": Fannish storytelling through new media. In K. Hellekson & K. Busse (Eds.), *Fan fiction and fan communities in the age of the Internet* (pp. 245–260). Jefferson, NC: McFarland.

Szadkowski, J. (2007, August 25). Resorts' MagiQuest actives slick fantasy. *Washington Times*, p. D01. Retrieved May 20, 2012, from http://www.washingtontimes.com/news/2007/aug/25/resorts-magiquest-activates-slick-fantasy/

Uren, T. (2007). Finding the game in improvised theater. In P. Harrigan & N. Wardrip-Fruin (Eds.), *Second person: Role-playing and story in games and playable media* (pp. 279–284). Cambridge, MA: MIT Press.

CHAPTER SIX

Collaborative, Productive, Performative, Templated: Youth, Identity, and Breaking the Fourth Wall Online

Shayla Thiel-Stern

With a few dozen keystrokes on her family's laptop at the kitchen table, 15-year-old Madeleine announces that she is married, she likes Katy Perry's new song and a comment about a poem that her younger sister posted on her wall (signed "Cya upstairs! Luv u!"), and that she is planning to dye her hair black this weekend.

Using Facebook and the extra half hour she has after school, Madeleine has quite publicly and iteratively articulated her identity to the world (or at least to her 634 Facebook friends). She is a fan of pop music, she has a sister with whom she gets along, and she is feeling a bit experimental (and perhaps rebellious) with regard to her hair color.

Approximately 20 minutes later, her mom comments about the hair color, "Can we talk about this when I get home?"

Madeleine is also the daughter of a fairly open-minded mother who uses Facebook at work.

The notion of adolescents using digital media to post information about both their current and potential personalities is no longer new. However, this process—which is directly linked to the theoretical concept of identity—has changed how many young people navigate adolescence. Identity is still negotiated in part through conversation and common cultural discourse, but thanks to digital media it is done so quickly, publicly, and often quite consciously. Adolescents are not merely consuming members of an online audience, as might have been thought in past mass media studies; they are media producers armed with mobile video cameras and knowledge of how to share recordings with their peers.

Facebook is only one of many tools available to youth who use the Internet or mobile devices, and it will not be the last social media tool used to articulate identity. Instant Messaging, or IM, an instant-relay chat tool popular in the late 1990s and early 2000s, allowed this through conversation and symbolism; IM is still used today, but is now a piece of other types of chat and even mobile texting. For a time, teens used blogging (especially through services such as LiveJournal) to record their thoughts and connect with one another; they still do this, although they might be using Blogger.com or Tumblr to publish.

In other words, the Internet is an established space where adolescents know they can connect and share with one another and where they can emphasize different aspects of their identity with the click of a mouse. As it has moved from a mass medium to a more social medium, the Web has become a more important vehicle for performance of identity than ever before. Goffman's (1959) theory of identity describes humans as social actors who perform identity along with everyone surrounding them all the time in simultaneous performance, with "front stage" being the aspects of a public performance of identity and "back stage" the hidden or less publicly performed pieces of identity. Although this conceptual definition of identity remains useful in understanding online and offline identity, the unique aspects of how adolescents use digital media and social networking sites could warrant revisiting this model. In addition to being a site for iterative, public identity negotiation, the Internet is a medium that encourages interacting with the audience—or what I call "breaking the fourth wall." On stage, three physical walls surround the actors; the fourth wall is the imaginary barrier at the front of the stage. The fourth wall concept enriches the dramaturgical concepts used by Goffman and others, to more fully frame how adolescents negotiate, articulate, and perform identity online today. This chapter will use these concepts as I consider the history of how media have portrayed adolescents' Internet use, and examine how identity can be negotiated and articulated through young people's production of discourse and cultural artifacts online. I theorize that identity construction has become collaborative and iterative— key components related to breaking through the fourth wall—but still rooted in dominant cultural discourse and increasingly templated. I briefly consider some of the methodological challenges of studying an audience that is both productive and aware of its audience, even as it conducts most of this production in the privacy of home.

Youth and Digital Media:
Mass Media Foster a Panicked Response

The mass media's representation of how adolescents use new media would be more accurately and typically described as *adolescents being used by* new media. In the early days of the Web, parents were warned about children accidentally happening upon pornography online. News reports using alarming language and spooky music delivered frightening narratives about young people entering chat rooms where sexual predators lurked and lured them into real-life meetings (Thiel-Stern, 2009). The best-known of these types of reports were taken up by the television newsmagazine *Dateline NBC*, which spawned a popular hidden-camera show, *To Catch a Predator* (2004–2008). Years later, news reports shed light on the growing phenomenon of online bullying. Hundreds of stories about cyber bullying have been published or aired between 2009 and the present (I am often interviewed by reporters about cyber bullying). Although these are important concerns and lines of conversation for parents, educators and young people, the media's representations of the Internet as a Wild West where youth are powerless victims is overhyped and overgeneralized.

Much of the news coverage of youth and the Internet is rooted in the discourse of moral panic. Stanley Cohen's classic definition of moral panic suggests it occurs when a "condition, episode, person or group of persons emerges to become defined as a threat to societal values and interests" (1972, p. 9). The media usually present an issue in a "stylized and stereotypical fashion" (p. 9). Cohen warned that although some panics may be forgotten, others have long-lasting repercussions, including legal and social policies; again, the Internet and the way young people use digital media are constantly up for discussion within government and among policy experts. Especially in its early days, the Web mystified most people who were not technologically inclined; ten years later, social networking sites had a similar effect amplified at first by generational differences. Critcher (2003) said media foster and generate moral panic especially when people depend on the mass media to explain unknown phenomena: "Since most people have no first-hand knowledge of deviants, they are reliant on the media. Understanding the role of the media becomes central" (p. 11). Cohen (1972) explained that even when media workers are not engaging in muckraking or crusading against an issue, "their very reporting of certain facts can be sufficient to generate concern, anxiety, indignation or panic" (p. 16). Thus, by choosing to focus newsgathering and reporting on how the Internet can be harmful, the news media can foster moral panic, and in the process establish a dominant discourse about the Internet as an unsafe space for young people.

Perhaps it is the relative ease and privacy with which young people use the Internet—often either alone in a private space in their home or hunkered down staring into the tiny screen of a smartphone—that feed the imagery of the Internet as an evil, impossible-to-stop villain. However, this ease and privacy also allow young people to feel safe in negotiating and navigating the often-complicated terrain of adolescence. The mutability of the digital makes it easy for audiences to shift identity and to enact different aspects of an identity online, even on the same profile page, several times a day.

Identity, Youth, and Studying Performance Online

In order to better understand how identity is negotiated and performed online, it is helpful to understand this process as productive and iterative, but also interpellated and templated.

Online Identity as Productive

Whereas educators and parents alike are often hesitant to consider young people's use of digital media at home or in the classroom a process of production, I suggest that even a simple conversation on Instant Messaging can be an act of cultural production. Most communication scholars agree that teens are not empty receptacles or passive audiences of media (as they might have been considered in the early flow models or strong effects theories of mass communication), but even those who agree that the youth audience is capable of actively negotiating meaning in a media text might hesitate to call the work of online audience members "productive." Arguably, updating one's whereabouts and preferences, sharing video and audio, and commenting upon other people's behaviors suggest activity. Updating and sharing online also produce a lasting record, or an artifact, that can be viewed and recalled. Although dominant media discourses often present the digital world as one where cyber bullying, sexual predation, and other ills abound, interactive media from Instant Messaging in the early 2000s to Facebook today have afforded young people an opportunity to productively articulate, construct, and negotiate identity in a way that was not possible before so many had access to the Internet or mobile devices. In doing so, they have created a wealth of online artifacts, from personal conversations to creative writing and images, that could, and should, be classified as productive.

This is not a new phenomenon. Since 1995, when the World Wide Web became accessible through the Mosaic and then Netscape browsers, young people have claimed the Internet as their own, out of earshot and view of parents and other authority figures (Thiel, 2005; Thiel-Stern, 2007). In the early days of the Web, tech-savvy teens learned how to build their own Web pages with their own personal preferences and imagery (Stern, 2004). Others used forums and online communities to communicate, usually about gaming and hobbies (Abbott, 1998; Tobin, 1998); they also used IM (primarily the AIM messenger client from AOL or Microsoft's MSN) to extend conversations with classmates and friends after the school day ended. With these tools of interactive media, adolescents not only connected and communicated with one another but they also broadcast aspects of identity through discourse with peers and others. Much of this discursive identity performance online was done through writing—in conversational text. In the early days of the Web and mobile devices, users had only alphanumeric keys at their disposal to make their points. Writers online appropriated symbols or combinations of letters, numbers, and punctuation marks to create emoticons, and to extend identity performance beyond what was available through text alone. These written conversational performances constitute a kind of *produsage*—a means for non-professionals to create or repackage an existing artifact to create an entirely new product for their own purposes and meanings (Bruns, 2005; 2006; Picone, 2011; Proulx, Heaton, Kwok, Choon, & Millette, 2011).

Hermes (2009) encouraged us to reimagine the contemporary digital media audience as an ad hoc cultural producer. Mazzarella (2007) and Kearney (2006) urged us to consider adolescents (girls, specifically) active producers of media instead of relegating them to the consumer realm as is so often done in popular media portrayals. *Produsage* has become increasingly more available to youth as new media tools have become more accessible and simple. Within a few years of the wide adoption of IM and mobile texting, blogging software enabled even non-tech-savvy teens the opportunity to build pages of their own, and to easily (and affordably) publish journal-like accounts of their lives. Soon after, social networking sites offered a template replete with digital photos and a fill-in-the-blank approach to expressing "likes" and "dislikes."

Online Identity as Interpellated

As the Internet and especially social media blur the boundaries between audiences and producers, we must consider not only the production and products themselves, but also the visibility of this creation and the ways in which identity is both articulated and hidden through the available tools (the templated social network and Web-site-building experience) as well as cultural norms (the adolescent who uses a sexy advertisement as her profile photo).

In research about adolescent girls and their use of Instant Messaging that I conducted between 2001 and 2004, I expected that tweens and teens would use IM as a vehicle of empowerment and produsage (Proulx et al., 2011), a means of speaking their minds or approaching a romantic interest without being under the critical gaze of their parents or peers. I envisioned an adolescent world very different from my own, where I sometimes felt awkward and afraid to express my feelings and passions because it meant I would most certainly blush or look like a geek, or both. After immersing myself in the research, interviewing and reading the IM conversations of girls aged 12–16 from various races and income levels over an extended period of time, I realized I had found a far more complicated online world than I imagined. First, adolescent girls certainly used IM to negotiate gender identity and sexuality through their conversations, the imagery they used to represent themselves, and the actual practices associated with IM (printing and keeping important conversations like diaries, copying pieces of conversations with one person and pasting them into a conversation with another, or not responding too quickly to messages to make it look as though they are busy conversing with all their other friends). Sometimes, these practices seemed to subvert gender norms (for example, the profanity-laden conversations from the girl who was a sweet, model student and whose parents assured me she never swore when they volunteered her for my study). More often, however, they fell comfortably within dominant stereotypes, from speaking about sex in a way that conjured pornographic imagery in which the primary recipient of sexual pleasure was the male, or performing the role of mother or caregiver to teenage boyfriends. However, these digital media practices were largely considered somewhat private—even by the adolescent girls who essentially knew better because they themselves would share their conversations with peers.

They also viewed IM as a safe space to enact and perform identity through private conversations with many different people. This practice sometimes seemed to resonate with Hall's (1996) interpretation of identity being in process and fluctuating, as well as Turkle's (1995) view of the online realm providing a space for identity play, but much of the discourse itself seemed more closely aligned with Althusser (1970) and interpellation. Even when considering the participants culturally productive in what they created through IM, it was difficult to move away from the notion that identity was being negotiated while the dominant ideologies—often perpetuated by the traditional mass media—lurked in the corner of the girls' eyes.

Early scholarly literature on identity and the Internet (Reid, 1991; Turkle, 1995) focused on how people could intentionally mislead one another or engage in identity play (regarding cultural markers of identity including race, ethnicity, age, and gender). However, early research on young people and the Internet often suggested something different: they viewed the Internet as a space where they could be more authentically themselves (Tobin, 1998) or have more "pure relationships" (Clark, 1998), and where they perceived their own presentation of themselves as fairly consistent (Lewis & Fabos, 2000; Thiel, 2005). Young people today often attest to the same. The increasingly visual nature of the Internet and the ability to Google one another contributes to this, but so does the notion that the Internet allows people to really be themselves and seek out other like-minded individuals. For example, the documentary *Growing Up Online* (Maggio & Dretzin, 2008) focuses on two teens who consider their online selves more real than their offline selves; one visited Pro-Ana sites created by people with eating disorders offering tips for how to stay thin and hide their disease, and the other posted artistic but explicit photos of herself on MySpace and became something of an Internet sensation, unbeknownst to her parents. Both are somewhat troubling stories, but statistics have demonstrated that most teens feel more comfortable expressing themselves online than offline, even as they become more savvy about maintaining privacy by using privacy settings to guard against parents or other authority figures checking up on them (Lenhart, Madden, Smith, Purcell, Zickuhr, & Rainie, 2011).

Moreover, the ability to negotiate gender identity through digital media can present an opportunity for marginalized or silenced teens to be heard and to act as cultural producers. Adolescent girls from varying races, ethnicities and classes today are prolific bloggers (Mitchell, Pascarella, deLange, & Stuart, 2010; Vickery, 2010). They create fan communities and write fan fiction, compose pages on Tumblr and other social networking sites (Bae, 2010; Warburton, 2010), and produce and post digital films (Kearney, 2006; Sweeney, 2008). Digital media allow them to turn the tables on corporate mainstream media and demonstrate that they are in fact able to represent themselves and produce media on their own terms (Mazzarella, 2005).

Still, even with the ability to articulate identity on one's own terms, the Internet does not offer absolute agency. Identity is still bound to cultural discourse related to acceptable cultural and social behaviors (Foucault, 1974). All people mark themselves or are marked by others as they exist within dominant cultural systems and ideologies, and frankly, cannot exist outside of these. Within these systems, humans believe they are in charge of how they behave, identify themselves, and function, even though it's largely an illusion based on a person's understanding of the system (Althusser, 1970). Rather, people articulate identity in part as a response to their being interpellated (or "hailed") by the ideological systems constantly surrounding them (Althusser, 1970, p. 162). This is evident in the sexy Abercrombie & Fitch advertisements a 14-year-old girl might post as her logo in IM or on Facebook instead of her own picture; it is also evident in the way that she might pose herself in a picture that looks remarkably like an Abercrombie and Fitch ad (Thiel-Stern, 2007).

Judith Butler's (1997) contention that gender identity is a repetitively performed construction based upon a person's understanding of dominant cultural discourses suggests hailing, too. Females and males perform what they interpret their gender to be, based upon what culture has taught them is the correct (in most cases, heterosexual) interpretation of gender and dominant understandings of masculinity and femininity. Moreover, Stuart Hall understands identity as "fragmented," "fractured," and a construction of a narrative that is comforting to an individual but always "in process" (1996, pp. 4–5). The ways that adolescents use digital media further emphasize how identity can be at once interpellated and in process through the iterative, interactive nature of the Internet itself.

Online Identity as Iterative

Identity construction for adolescent audiences has become increasingly iterative, even collaborative. In IM, chat, or Facebook postings, the conversation itself often acts as both a diary and a space for comment. Young people negotiate issues from body image to sexuality to schoolwork conversationally, and often they will save these conversations by printing them out and storing them in a safe place (Thiel-Stern, 2007) or simply return to them using Facebook's "See Friendship" option. Although these exchanges are often about events and experiences, they are just as often about feelings, preferences, and a negotiation of the social landscape, and all of these are linked to identity and its performance. These performances also happen to include a peer offering affirmation (responding by hitting the "like" button or leaving a comment) of this piece of individual identity.

As this adolescent audience has become even more productive through its use of Facebook, Twitter, Google sites, and the myriad mobile apps and tools available today, it continues to do so in a mediated environment that does not make identity articulation as playful or empowering as I had envisioned when I set out in my early research. Perhaps this is because adolescents no longer view these communication tools and apps as enabling privacy in any way. In fact, being watched by and connected to others is a major part of the attraction of using social media. For example, in some recent research on adolescents and Facebook, I noticed that a 16-year-old male participant constantly posted status updates that seemed to involve only five of his 1,570 Facebook friends. When I asked him about it, he realized he had not been consciously doing it, and that the updates were all inside jokes among him and his friends from the football team. In engaging in this kind of practice, he seemed to be asserting his social status and (masculine) athletic ability, and creating something akin to the "cool kids at the lunch table" effect. These status updates certainly projected identity, and when the friends he mentioned (and many others who were not mentioned), liked or commented on the post in any way, it seemed to not only increase his visibility in his larger community of friends but also reinforce the identity that he projected. Identity in this way is as iterative as it was in IM conversations, but it also is public, and as such, valued by both the football player and his peers online.

The Internet expands upon the ways in which identity can be performed today. In addition to being iteratively produced through conversations and status updates, identity performances online are also tied up in visual imagery and artifacts of popular culture. Posting a YouTube video to a profile page on a blog or social networking site can become an act of identity articulation in which the poster expects an audience to comment and thus acknowledge their existence and choice of posting. Sometimes this choice might be a way of speaking back to peers (such as 14-year-old Kate's status "rumors tell me things i never knew about myself...how interesting (:"). Sometimes it might be an attempt to show a more rebellious side (such as the photo posted by the afore mentioned Madeleine after she dyed her hair from brown to black). In both of these cases, the girls received "likes" in the double digits along with a bevy of positive comments from their friends, reinforcing these particular aspects of their identity. Now more than ever with social media, identity articulation is immediate and instant, but ever shifting. It is visual and textual but also mediated and rooted in dominant cultural discourse that dictates appropriate social norms. Its users are still both conscious and unconscious of the representation of self as they shift among types of digital media.

Facebook founder Mark Zuckerberg has been quoted as saying, "You have one identity. The days of you having a different image for your work friends or co-workers and for other people you know are probably coming to an end pretty quickly. Having two identities for yourself is an example of a lack of integrity" (Kirkpatrick, 2009, p. 199). Despite his insistence on consistency of identity, the truth is that identity is still constantly shifting and performed, even on Facebook. Increasingly, social networking sites allow their users to perform different aspects of identity for different audiences. For example, a site allowing users to define different clusters of their connections (audiences) and share, for example, only work-related information with some people and personal information with others, allows—and even encourages—identity to be performed differently for different audiences.

Digital media in the not-so-distant past afforded a relatively high level of both anonymity and literal disembodiment. The news media in the 1990s and early 2000s depended on this specific conceit in their narratives because in a chat room, anyone—even a child molester—could pose as anyone else. There were no profiles, photos, or consistent "identities" (in the Zuckerbergian definition) attached to humans in chat rooms or other virtual spaces until many years later. In addition to people voluntarily posting artifacts from their lives on social networks, Google allows audiences to search and organize that information on one page in response to a single search on a name—and to click "Images" if they want to *see* what they've found.

However, adolescents often get a pass here. Few have amassed an online presence or a bevy of images, as adults have through years of working for companies that post employee photos and biographies on corporate Web sites, or by posting on listservs about their favorite rock bands (these are my two main Googleable aspects of identity). Adolescents can easily post whatever photos they want to represent themselves online using social media, from a self-portrait framed in exactly the way they prefer to a corporate logo for a clothing brand that they like to a picture of Santa Claus. (One 15-year-old from my own research used all three of these within a single hour of time while chatting online.)

In addition to becoming iterative, identity as performed in the age of social media has also become increasingly public. In other words, the audience has an audience, and the audience—especially teenagers—is increasingly becoming aware of this (Thiel-Stern, forthcoming). Goffman's (1959) notions of front stage and back stage performances focused on the former as the idea that humans perform social roles visibly and in front of an audience and the latter as behaviors reserved for close friends and family members. However, Goffman's ideas about back stage might be understood differently in the context of social media because often even private behaviors and feelings are performed for an audience; on social networking sites, the audience is always there, even if the performer is alone and silent, sitting in front of a computer screen or iPhone. This is where the concept of "breaking the fourth wall" might be a helpful framework for understanding how adolescents perform identity today.

Identity Moving From Back Stage Through the Fourth Wall

In addition to being aware of their ability to represent themselves or perform identity, adolescents also seem aware of something akin to the invisible wall between performer and audience, and tend to break this fourth wall in their daily performance of identity in social media. In fact, young people who use social media often are so aware of identity performance and audience that in essence, they constantly break the fourth wall, and in many cases, relish that space.

Walls are a useful metaphor when discussing adolescents and their use of new media today. Most teens are well aware that they are judged as much by the pop culture preferences they post on Facebook as by posts they leave on their own and others' walls. Bedroom walls used to be a primary space where adolescents could articulate identity through posters, paint, and décor that they had chosen (Steele & Brown, 1995), but social media give them a far more public stage to list their favorites, post song lyrics and photos or video performances of their favorite actors and artists, and write or say whatever is on their mind at a given moment. Furthermore, it is not an imagined, potential audience (such as classmates who might, or might not, drop by for a visit) but an audience they have chosen specifically, which more often than not is online at least once a day taking note of them. Services such as Facebook and Google+ tell users who else is online at a given moment, so it's simple for young people to realize exactly who is in their audience at any given time and cater their messages accordingly. Because they choose who they add to their networks (and use privacy settings or circles to determine who in those networks sees certain information), the performers become all the more self-aware and perhaps calculating in how they address their various audiences. This is very much like an actor breaking the fourth wall— whether it is Woody Allen telling the camera what an idiot a certain Media and Culture professor is in *Annie Hall* (a scene in which Woody brings out Marshall McLuhan to tell the idiot professor that his ideas are wrong) or Shakespeare's Hamlet addressing the audience directly in a soliloquy. It indicates the same calculated awareness and the same address of audience, except it is made through a status update rather than a script. This wall may or may not function both ways. Much as with a film or play, the audience can choose to actively respond or interact to the posts or simply observe.

After breaking the fourth wall, stage performers do not necessarily expect a response, and film actors cannot get one. However, social media users
often need immediate affirmation—even a simple action such as friends hit
ting the like button to acknowledge they enjoyed the post. In fact, a 19-year-
old college freshmen recently told me that she will delete any posts on Facebook that do not garner an immediate response from her friends. She said she
admitted this to her girlfriends, and they responded that they did the same
thing. Perhaps the usability of sites plays a part here as well. It becomes as
easy to delete or edit as to post.

Identity as a Templated Experience

Despite the potential for creative identity articulation, it must be noted that
adolescents' construction of identity online has become increasingly
templated in recent years. In the early years of the Web, kids had to
understand HTML and have server space to upload their pages to in order to
publish personal Web sites. They were limited not only by knowledge of the
technology but also by income, because hosting a site could involve paying a
fee. In the 1980s, 1990s, and early 2000s, the Internet required more of its
young users—technical acumen, the drive to create something for the public,
a computer, and the money to host. Today, far less is required because so
many sites have developed WYSIWYG tools that make designing a breeze
and hosting free. Social networking sites allow users to create pages of their
own that function in the same way that early Web pages functioned for
adolescents—but with far less headache and an audience that they
specifically choose through the act of "friending." On the one hand, social
networking sites and free Web media tools such as Google Sites and
podcasting software have provided an inclusive experience by allowing teens
with little technical expertise in technology to participate in the creation of
digital media. On the other, they rob young people of technical acumen and
creative license, as well as the ability to imagine beyond the usable but rather
boring white and blue template they have become accustomed to seeing on
Facebook.

When theorizing, we should consider what it means that young people
often articulate identity through these simple tools. Referring to the earlier
discussion of whether identity is still based on understandings of hegemonic
and accepted ideals, it is possible that identity is very much an act of cultural
hailing. However, identity is also performed in part based on the amount of
space to upload one's hobbies, preferences, and relationship status, and often
limited to a profile photo or videos that are short enough to post to a site.

Identity in the digital age may likely be performed in both a cultural and an online template. Furthermore, within both the cultural and online template, we must consider how and to what degree identity performance is based upon the assumption of an audience and the interaction with that audience, despite the methodological challenges associated with studying young people online.

Methodological Challenges

Public, iterative identity negotiation among adolescents is only part of the story. After all, public identity often differs from private identity, as Goffman (1959) pointed out in his conceptualization of front and back stage performance. The way that adolescents behave online among peers might be very much at odds with their intentions as they post on laptops from the comfort of their own kitchen tables. I discovered just this in my study of IM use by adolescents; they often engaged in private conversations about religion and sexuality via IM but seemed much cooler and more ambivalent about such matters in their in-person conversations with me, or even prone to fits of giggles in our group interviews. This suggests that we should consider new methodologies and research paradigms in our study of youth audiences and digital media. Press and Livingstone (2006) have acknowledged the challenge of studying the public/private and audience/producer dynamic because youth's Internet use is largely private, the engagement is often personal or transgressive, and "recording and interpreting an evening's surfing or chat" is tricky (p. 187). Livingstone considered this position in later research of adolescents' use of social networking sites (2008), in which she interviewed 16 British teens in their homes as they demonstrated how they used MySpace, Facebook, and Piczo (a blogging community). That article demonstrated how gathering rich data, which account for the shifting dialectic between audience and producer, can better yield theoretical insight.

Classroom studies of adolescents and digital media can provide a rich understanding of youth, new media, and identity because they enable researchers to observe the creation of an online product, view the product itself, and ask questions about the student's thoughts on the product. In participant-observation work with a public high school media studies class in Minneapolis, I used this inclusive approach to understand how students articulated and negotiated identity through the digital media products they created in school. Although the students had access to the Internet at school and in community libraries, many of them did not have access to the Web or smartphones at home, which made the study of their digital media use and production at school important. Abdi, a 17-year-old Somali boy without access at home, used his time in class to produce a Google site and a blog that included various entries about NBA basketball, his excitement about graduation, and a link to a YouTube video that he made in an after-school program about gun violence. Although he starred in the short clip in which he was given a gun by a gang member, he explained that he had never owned a gun nor been in a gang. I had not questioned him either way, but he felt it was important to clarify this to me in person. However, he said that if his audience questioned the performance, they could use the comment section and he would respond and explain. Already, it seemed, he anticipated questions related to his identity, both online and offline. It was as if Abdi expected the need to break that fourth wall between his performance and his audience, and he already knew exactly where on his site he could do so should the need arise. Perhaps his own media literacy skills were advanced and sophisticated enough to lead him to this realization, but I also see his exploitation of his ability to break the fourth wall as a common bond among other adolescents. They may be developing the ability to analyze audiences in a way that previous generations found either elusive or unnecessary.

Abdi's anticipation of public response and his own public answer to this potential response is an interesting consideration in the study of youth and digital media, and it suggests that identity as a concept has become all the more complicated. Feeling in the back of your mind that hundreds or thousands of potential viewers could be watching and waiting to question the status updates or amateur videos that you post certainly would affect what you choose to share and how you choose to share it. Adolescents might be counting on that conversational give-and-take to make decisions about how they will represent themselves online and how they navigate adolescence itself. Given this premise, some might worry that a generation of young people will not learn to develop and negotiate identity independently as they enter adulthood. This, like many people's understanding of youth and new media, could be overblown and may possibly verge on moral panic, but it is a vital line of questioning in terms of the sociological impact of digital media on youth. Yes, both Goffman's (1959) and Butler's (1997) theoretical understanding of identity as a performance are still in play, but these constructs of identity are amped up by the immediacy and hyper-social nature of digital media today. Perhaps the process becomes less about public identity articulation, and more about public identity production, in a space where the fourth wall is constantly, consciously, being breached.

References

Abbott, C. (1998). Making connections: Young people and the Internet. In J. Sefton-Green (Ed.), *Digital diversions: Youth culture in the age of multimedia* (pp. 84–105). New York, NY: Routledge.

Althusser, L. (1970). Ideology and ideological state apparatuses. In *Lenin and philosophy and other essays* (B. Brewster, Trans.) (pp. 121–250). New York, NY: Monthly Review Press.

Bae, M. (2010). Go Cyworld! Korean diasporic girls producing new Korean femininity. In S. R. Mazzarella (Ed.), *Girl wide web 2.0* (pp. 91–116). New York, NY: Peter Lang.

Bruns, A. (2005, March 11). *Some exploratory notes on produsers and produsage.* Retrieved February 14, 2012, from http://snurb.info/node/329

Bruns, A. (2006). *Blogs, Wikipedia, Second Life and beyond: From production to produsage.* New York, NY: Peter Lang.

Butler, J. (1997). Performative acts and gender constitution: An essay in phenomenology and feminist theory. In K. Conboy, N. Medina, & S. Stanbury (Eds.), *Writing on the body: Female embodiment and feminist theory* (pp. 401–418). New York, NY: Columbia University Press.

Clark, L. S. (1998). Dating on the Net: Teens and the rise of "pure" relationships. In S. Jones (Ed.), *Cybersociety 2.0* (pp.159–183). Thousand Oaks, CA: Sage.

Cohen, S. (1972). *Folk devils and moral panics: The creation of mods and rockers.* London, England: MacGibbon & Kee.

Critcher C. (2003). *Moral panics and the media.* Buckingham, England: Open University Press.

Foucault, M. (1974). *The history of sexuality.* New York, NY: Vintage Books.

Goffman, E. (1959). *The presentation of self in every day life.* New York, NY: Anchor Books.

Hall, S. (1996). Who needs 'identity'? In S. Hall & P. du Gay (Eds.), *Questions of cultural identity* (pp. 1–17). Thousand Oaks, CA: Sage.

Hermes, J. (2009). Audience studies 2.0: On the theory, politics and method of qualitative audience research. *Interactions: Studies in Communication and Culture, 1*(1) 111–125.

Kearney, M. C. (2006). *Girls make media.* New York, NY: Routledge.

Kirkpatrick, D. (2009). *The Facebook effect: The inside story of the company that is connecting the world.* New York, NY: Simon & Schuster.

Lenhart, A., Madden, M., Smith, A., Purcell, K., Zickuhr, K., & Rainie, L. (2011). *Teens, kindness, and cruelty on social network sites.* Retrieved Nov. 21, 2011, from Pew Internet and American Life Project Report: http://www.pewinternet.org/Reports/2011/Teens-and-social-media/Part-3/Deciding-not-to-post.aspx

Lewis, C., & Fabos, B. (2000). But will it work in the heartland? A response and illustration. *Journal of Adolescent & Adult Literacy, 43*(5), 462–469.

Livingstone, S. (2008). Taking risky opportunities in youthful content creation: tTeenagers' use of social networking sites for intimacy, privacy and self-expression. *New Media & Society, 10*(3), 393–411.

Maggio, J. (Producer), & Dretzin, R. (Director). (2008). Growing up online [Video file]. *Frontline.* Boston, MA: WGBH Boston and Public Broadcasting System. Retrieved Feb. 17, 2012, from http://www.pbs.org/wgbh/pages/frontline/kidsonline/

Mazzarella, S. R. (2005). Claiming a space: The cultural economy of teen girl fandom on the Web. In S. R. Mazzarella (Ed.), *Girl wide web: Girls, the Internet, and the negotiation of identity* (pp. 141–160). New York, NY: Peter Lang.

Mazzarella, S. R. (2007). How are girls' studies scholars (and girls themselves) shaking up the way we think about girls and media? In S. R. Mazzarella (Ed.), *20 questions about youth and the media* (pp. 253–266). New York, NY: Peter Lang.

Mitchell, C., Pascarella, J., De Lange, N., & Stuart, J. (2010). We wanted other people to learn from us: Girls blogging in rural South Africa in the Age of AIDS. In S. R. Mazzarella (Ed.), *Girl Wide Web 2.0* (pp. 161–182). New York, NY: Peter Lang.

Picone, I. (2011). Produsage as form of self-publication: A qualitative study of casual news produsage. *New Review of Hypermedia and Multimedia, 17*(1), 99–120.

Press, A., & Livingstone, S. (2006). Taking audience research into the age of new media: Old problems and new challenges. In M. White & J. Schwoch (Eds.), *Questions of method in cultural studies* (pp. 175–200). Oxford, England: Blackwell.

Proulx, S., Heaton, L., Kwok Choon, M., & Millette, M. (2011). Paradoxical empowerment of *produsers* in the context of informational capitalism. *New Review of Hypermedia and Multimedia, 17*(1), 9–29.

Reid, E. (1991). *Electropolis: Communication and community on Internet Relay Chat* (Unpublished honor's thesis). University of Melbourne. Retrieved Nov. 12, 2011, from http://eserver.org/cyber/reid.txt

Steele, J. R., & Brown, J. D. (1995). Adolescent room culture: Studying media in the context of everyday life. *Journal of Youth and Adolescence, 24*(5), 551–576.

Stern, S. R. (2004). Expressions of identity online: Prominent features and gender differences in adolescents' World Wide Web home pages. *Journal of Broadcasting & Electronic Media, 48*(2), 218–243.

Sweeney, K. (2008). *Maiden U.S.A.: Girl icons come of age.* New York, NY: Peter Lang.

Thiel, S. (2005). "IM Me": Identity construction and gender negotiation in the world of girls and instant messaging. In S. R. Mazzarella (Ed.), *Girl wide web: Girls, the Internet, and the negotiation of identity* (pp. 51–68). New York, NY: Peter Lang.

Thiel-Stern, S. (2007). *Instant identity: Adolescent girls and the world of instant messaging.* New York, NY: Peter Lang.

Thiel-Stern, S. (2009). Femininity out of control on the Internet: A critical analysis of media representations of gender, youth, and MySpace.com in international news discourses. *Girlhood Studies, 2*(1), 20–39.

Thiel-Stern, S. (Forthcoming). 'This is not an audience': Another reconsideration of audience studies in the interactive media environment. In R. E. Parameswaren (Ed.), *Companion to Audience Studies.* New York, NY: Wiley-Blackwell.

Tobin, J. (1998). An American otaku (or, a boy's virtual life on the Net). In J. Sefton-Green (Ed.), *Digital diversions: Youth culture in the age of multimedia* (pp. 106–127). New York, NY: Routledge.

Turkle, S. (1995). *Life on the screen: Identity in the age of the Internet.* New York, NY: Touchstone.

Vickery, J. R. (2010). Blogrings as virtual communities for adolescent girls. In S. R. Mazzarella (Ed.), *Girl wide web 2.0* (pp. 183–201). New York, NY: Peter Lang.

Warburton, J. (2010). Me/Her/Draco Malfoy: Fangirl communities and their fictions. In S. R. Mazzarella (Ed.), *Girl wide web 2.0* (pp. 117–138). New York, NY: Peter Lang.

CHAPTER SEVEN

Psychoanalysis and Digital Fandom: Theorizing Spoilers and Fans' Self-Narratives

Matt Hills

It could reasonably be argued that the psychoanalysis of media fandom—let alone that of online/digital fandom—has been a marginal aspect of fan studies. From Henry Jenkins' seminal work *Textual Poachers* (1992) and onward through various waves of scholarship identified by Gray, Sandvoss and Harrington (2007), psychoanalysis has had an awkward place in the field, despite some attempts to reclaim or stake out its centrality (Harrington & Bielby, 1995; Hills, 2002). Overviews have analytically separated out Kleinian work on fans (Hills, 2002; Sandvoss, 2005), approaches based on the writings of Laplanche and Pontalis (Hills, 2002), and a turn to Winnicottian object-relations (Hills, 2002; Sandvoss, 2005), as well as examining broadly Freudian work on fans' fantasies (Sandvoss, 2005). However, the relative absence of psychoanalysis in fan studies can, in part, be attributed to ethical concerns. As John Fiske noted quite some time ago: "psychoanalytic...theories...allow the theorist a privileged insight into the experiences of their subjects that is not available to the subjects themselves" (1990, p. 90).

Psychoanalysis thus seemed not to fit with an emergent orthodoxy of giving voice to fans who had previously been negatively stereotyped and pathologized in both prior media/cultural theory and the media itself. And given the movement in media studies from psychoanalytically oriented film studies to ethnographic and cyber-ethnographic work, there has also been "no significant place for psychoanalytically-inflected research in the study of...virtual communication" (Bailey, 2005, p. 38). Despite book-length studies of female fandoms online (Bury, 2005), adolescents' online fan fiction

(Black, 2008), and *Digital Fandom* (Booth, 2010), psychoanalysis has remained largely beyond the pale in (online) fan studies.

This is more than a little surprising, however, given that although "specific assumptions regarding what and who a fan is may vary, it seems safe to assume that we can associate fandom with a particular form of emotional intensity or 'affect'" (Sandvoss, 2005). Likewise, although I have described fandom as "always performative" (i.e., it varies in discursive practices and cultural contexts) (2002, p. xi), and although Rhiannon Bury has described "the fan" as an "invisible fiction" or discursive construct (2005, p. 205), there remains a general agreement—in line with Sandvoss's observation—that fandom is marked by "intense emotional investment" (Bury, 2005, p. 208) and "affective play" (Hills, 2002, p. 64). As such, it would seem to be a strong candidate for psychoanalytic exploration.

However, a further difficulty is posed by the equation of fandom with strong emotion: this has frequently formed a part of fandom's pathologization. As well as claiming to understand fans in ways that they themselves cannot, psychoanalysis has run the risk of becoming caught up in fandom's problematic cultural devaluations, representing fan practices as a "psychopathology of everyday life" (Freud, 1901/1976). For instance, writing about a specific online fan community, Julian Hoxter (2000) made this unnerving slip between discussing fans and children:

> It is…doubtful whether Internet fan networks currently function as truly receptive containers which can understand and return and therefore strengthen the fan's (infant's) own capacity to contain. However, that is not to suggest that they perform no useful purpose. On the contrary, they clearly work at the level of providing basic comfort and security simply through the recognition of commonality of experience. (p. 185)

To be clear, Hoxter was not actually discussing children here. Yet in applying the psychoanalytic theories of Melanie Klein and Wilfred Bion, he appears to link online fan activity surrounding *The Exorcist* with infants' (psychodynamic) processes. And in case we assume that such a pathologizing maneuver has, by now, been thoroughly done away with, Jan Jagodzinski has carried out a broadly Lacanian analysis of what he terms the "fan(addict)" and their "pathological overidentification" (2004, pp. 58, 60; see also Jagodzinski, 2008, p. 197). The bracketed term again gives the game away, suggesting that Fiske's ethical objections to psychoanalysis may continue to carry some force. Not unrelatedly, Justin Smith has also recently argued that my own *Fan Cultures* (2002) fails to consider that "the extremes of fan devotion" could be "other than normal. Could it not be that fans who establish what might be termed an unhealthy fixation upon a certain cultural object have not as infants effectively decathected transitional objects[?]" (Smith, 2010, p. 186). Whether fandom is unhealthy, pathological, or fails to facilitate emotional growth and development, these psychoanalytic notions of fixated, over-emotional (online) fans hardly help us to unpack, challenge and contest cultural notions of fan abnormality (see also Bainbridge & Yates, 2010).

In fact, where the use of psychoanalysis in cultural theory is concerned, discourses of pathology are seemingly never far away. Ian Craib has inverted the usual terms of debate by arguing that sociological approaches to identity—lacking any concept of the unconscious and typically focused instead on cognitive processes of agency—are themselves normotic:

> all too often our understanding assumes that we live in a world of people who…in Christopher Bollas' sense, are "normotic." …we assume that cognition dominates people's lives, that we only have ideas and those ideas come to us from the outside, from the social world; we take them in and act on them. There are sometimes references to "interpretive processes" but no real exploration of what they involve. (Craib, 1998, p. 1–2)

Craib (2001, p. 189) borrowed this idea from Bollas's psychoanalytic clinical work, implying that theories of cultural identity construction, and audiences' cognitive agency, are akin to normotic illness:

> the normotic is interested in facts.… It is part of a personal evolution in which he unconsciously attempts to become an object in the object world. To collect facts is ultimately to be identified with that which is collected: to become a fact in one's person. It is truly reassuring to become part of the machinery of production. (Bollas, 1987, p.138)

In *Psychoanalysis and Culture*, Rosalind Minsky made a similar accusation, arguing that "the recent popularity of culturalist theories which argue that identity is an exclusively social or cultural construction, may be a reflection of a widespread unconscious need to feel we are in control" (1998, p. 16). According to such logic, the denial of the unconscious in media/cultural studies stems from a need for (theoretical) mastery, and/or from an evacuation of interior, subjective life in favor of "abnormal normality" where subjective states and experiences are analyzed as if they are purely objects in the world (Bollas, 1987, p. 156). According to Minsky and Craib, it would be feasible to describe Henry Jenkins' (2006) approach to online fan practices of spoiler gathering as "normotic," given that Jenkins analyzes fans' networked information-gathering without focusing on any sense of fans' interior, subjective lives. Instead, "collective intelligence" (drawing on Pierre Levy's term) looks like a "reassuring…machinery of production" in Jenkins' depiction, within which fan selves can become identified with facts-as-objects, thus themselves becoming objects in the world. However, it seems extremely unhelpful to import the languages of pathology into academic debate. Rather than mischaracterising Jenkins' work, in the next section I want to consider how specific fan practices surrounding spoilers might be psychoanalytically theorized without falling back on discourses of pathology to characterize either online/digital fans or prior theorists.

Psychoanalyzing Spoilers:
Time, Ontological Security, and Creativity

At this point, it is necessary to offer a definition of the spoiler. Essentially, spoilers are bits of narrative information that circulate within fan cultures prior to the official mass availability of the narrative concerned (film, TV episode, game). Distinct from rumor or speculation, spoilers are information framed by, and occurring within, the pre-production or production process of a given media text. They are assumed to have industrial validity and veracity, whereas rumors are not tied to a specific production.

Spoilers can have a variety of sources—some are unofficial or fan-sourced, and some are official, appearing in promotional paratexts such a trailers, or industry materials such as agency CVs. The fact that unofficial spoilers are "viewer-created paratexts" (Gray, 2010, p. 143) means that they can often be strongly opposed by producers (who take them to represent a form of damage to the commercial value of a property). At this level, spoiling "is…a contest between the fans and the producers, one group trying to get their hands on the knowledge the other is trying to protect" (Jenkins, 2006 p. 43). Somewhat ironically, however, producers view official spoilers as enhancing brand value; they are about hyping a product, and so are professionally adjudged to properly control what is narratively revealed. Unofficial fan spoilers allegedly lack fine-tuned judgment and sufficient quality control. It isn't the spoiler per se that is industrially opposed in this "info-war" (Hills, 2010, p. 72), then, but rather its unofficial informational scope, scale, and lack of professional propriety. Unofficial spoilers threaten producers with a loss of mastery and control over the presentation of what they consider *their* narratives. And unlike official spoilers (with the exception of deliberately misleading production information), the veracity of an unofficial spoiler can only ever be deferred, because they can only become definitively true retrospectively, i.e., post-broadcast. In order for fake information to be sifted out or rejected, unofficial spoilers thus require careful fan deliberation and analysis, as well as careful evaluation of spoiler sources.

Spoilers can also be differentiated by the narrative significance of their information: revealing a character's name may be relatively unimportant (though even this is arguable, given that the character name could have significance in the wider franchise mythology, or could mark a returning figure). In contrast, revealing key plot twists and narrative outcomes are far greater spoilers. Varying in both degree and kind, spoilers are so-called because they are deemed to have the potential to spoil enjoyment of a storyline.

Henry Jenkins has pointed out in *Convergence Culture* that the term "spoiler" has at least one origin in Internet discourse, because fans who had already seen a TV episode in one U.S. time zone developed the netiquette of labelling their online discussions with so-called "spoiler warnings" in order that those who had not yet viewed the episode wouldn't read detailed analysis of it by accident (Jenkins, 2006, p. 30). In effect, the very concept of spoilers is therefore produced through "just-in-time" fandom (Hills, 2002, pp. 178–179), whereby fans favor discussing new episodes of their beloved TV shows in detail immediately after broadcast—also meaning that they will tend to do so before other (fan) audiences in different time zones, and different broadcast territories. Digital fandom therefore encounters a conflict between broadcast temporalities and online temporalities; for fans to participate in fandom's just-in-time informational tendencies, they need to have seen a TV episode on its worldwide premiere (or very shortly after, via file-sharing). This collision between new and old media means that, in fact, spoilers can be bits of narrative information circulating after a text's national availability, but prior to its international circulation. For this reason, online fan communities tend to be stratified along national lines—with different discussions following different national broadcast times. Where SNSs cannot be easily nationally stratified, e.g., following people on Twitter, there is a greater risk of unwanted exposure to spoilers.

Fans are thus required to accept a specific national broadcast as the de facto premiere and textual point of origin, meaning that if they wish to be involved in online discussion they must accept the possibility of being spoiled, or must illegally download in order to be able to participate in the fan community at times of peak discussion, debate and activity. As Robert Samuels has reminded us, "[w]hile it is easy to celebrate…the rise of the consumer as a producer of media, we need to pay attention to how this…narrative hides the synchronic layering of new media on top of old media" (2010, p. 33). Spoilers offer just one such layering.

Although spoilers have their origin in fan netiquette and just-in-time fandom, once the phenomenon had become subculturally recognized, spoilers then started to become a type of information actually sought and prized by sections of fandom, coming to mean pre-release/pre-broadcast information. For the "spoilerholic" (Cromarty & Lewin, 2011, p. 68) spoilers are a form of currency in both the temporal and axiological sense of that term: they represent information acquired as soon as is humanly possible—ahead of industrial regimes of publicity/promotion where unofficial spoilers are concerned—and they have a fan-cultural value, representing breaking news in the 24/7 rolling news informational economy of digital fandom.

Chuck Tryon has observed that the concept of just-in-time fandom needs to be complicated, because it fails to address the fact that fan blogging, micro-blogging, and posting happen online long before a text's release, meaning that fan anticipation and debate are stretched across a far more extensive time frame, with fan activities not being so clearly bound by industrial "spatiotemporal relationships" (Tryon, 2009, p. 136). His point is sound, and I would argue that the importance of breaking spoilers indicates that the informational economy of much digital fandom is fiercely attuned to a nowness and immediacy of knowledge. Spoilers may "work to dissolve the text's embeddedness with a relatively fixed set of discourses, as well as from the normal rhythms of the network television schedule" (Bailey, 2005, p. 179), but they do so by re-embedding narrative information precisely within a discursive practice of breaking news, and a hunger to find out as soon as is possible—especially ahead of other audiences, and even other fans. Spoilers indicate that just-in-time fandom's rhythms and temporalities have been generalized within sectors of 24/7, always-on digital fandom. It is no longer simply the broadcast premiere that generates massive spikes in online fan debate and activity, but also the acquisition of unofficial spoilers as well as the official provision of spoiler information.

The double articulation of spoiler currency—that is, information that is of intense value at its moment of acquisition, but which decays rapidly in value as it is superseded or becomes widely known—indicates that spoilers possess "communal, rather than individualistic, modes of reception" (Jenkins, 2006, p. 26). As a result, for Henry Jenkins "spoiling is collective intelligence in practice" (2006, p. 28), working to unite fan communities (or subcommunities) in the collective pursuit and accumulation of spoiler knowledge. And in *Digital Fandom*, Paul Booth similarly emphasized this communal orientation (2010). The communal pursuit of timely information can lead to fan hierarchies whereby certain contributors establish good online reputations (frequently providing spoilers that are later proven to be accurate), and then tend to "dump data with no explanation about how they got it, essentially cutting the plebians out of the process and constructing themselves as experts who are to be trusted at face value" (Jenkins, 2006, p. 39). Spoiling therefore creates its own elite fans, assumed to have insider knowledge and contacts, as well as raising the issue of:

> whether, within a knowledge community, one has the right to *not* know—or more precisely, whether each community member should be able to set the terms of how much they want to know and when they want to know it. [Pierre] Levy speaks about knowledge communities in terms of their democratic operations; yet the ability to dump information out there without regard to anyone else's preferences holds a deeply totalitarian dimension. (Jenkins, 2006, p. 55)

Jenkins typically celebrates collective intelligence as empowering for digital fandom (2006, p. 29), but he remains alert to the elitist and totalitarian dimensions of spoiling, where this can still very much live up to its name. Spoilers, then, are a kind of fan cultural production linked to specific concerns: they can threaten to impinge on fans' autonomy and enjoyment when they are unsought, and they can threaten to undermine producers' autonomy and control over *their* show.

I would say that spoilers are productively open to psychoanalytic interpretation because, unlike fan fictions or fan readings surrounding a text, spoilers centrally pose emotional questions of anxiety, trust, and control. Indeed, Jenkins cited Emily Nussbaum, writing in the *New York Times* on the subject of spoilers: "It's an odd wish—for control of the story, for the chance to minimize your risk of disappointment" (in Jenkins, 2006, p. 55). Nor has this anxiety-reducing potency of spoilers been missed by other scholars. Jonathan Gray and Jason Mittell noted how, in their survey of *Lost*'s spoiler devotees, "one fan wrote that he used spoilers to avoid investing his attention in relationships or characters that are doomed. ...As one respondent...[suggested:] "It's like reading a book and then watching the movie even when you know the ending" (as cited in Gray, 2010, p. 150). According to these accounts, spoilers can play particular (and I would add, psychical) roles within individual fans' emotional engagements. Thinking about spoilers as a containment of fan anxiety means complicating the community-oriented interpretations of Booth (2010) and Jenkins (2006). And as Mirko Tobias Schäfer has pointed out:

> Jenkins emphasizes the community aspects, the mutual understanding and genuine interest in each other's productions...but there are other user activities unfolding in the extensions of the cultural industries that revolve around different dynamics, and do not show tightly knit social relations or community identity. (2011, p. 43)

Spoilers may, though, have both community aspects *and* an anxiety-shielding psychical importance for individual fans going beyond community identity. The comparison with watching a movie of a book that's already been read is evocative, because it raises the implication that spoilers might be akin to a sort of re-reading. Gray and Mittell have "hypothesized that spoiler fans might enjoy spoilers because they preferred to watch in-the-know and were more comfortable with seeing the known than the unknown" (as cited in Gray, 2010, p. 149), although their small-scale study found the evidence on this to be inconclusive. Meanwhile, Paul Booth speculated that the "pleasure[s]...fans derive from spoilers come from experiencing the plot as known" (2010, p. 111).

Unlike spoilers, re-reading has already been psychoanalyzed, albeit problematically. Karen Odden argued that rereading is characteristic of children's engagement with texts—making no mention of fans—and suggested that "the reread world is reassuringly unresponsive to our own" inner world, giving it an anxiety-reducing quality of trustworthiness and reliability as an object (1998, p. 140). For Odden, rereading is not something that adults would need to do:

> As adults, we have more autonomy and power over our lives than we did as children; we have more power to remove ourselves from stressful situations and unfulfilling relationships; (it is hoped) we have found the world to be a place that is "safe enough" and we are more at ease with ambivalences and changes.... This in and of itself provides a security that we do not need to find in [rereading] novels. (1998, p. 149)

If spoilering is akin to rereading, then on Odden's account this provides a sense of security for the fan self. But this interpretation again relies on equating fans with children, amounting to a pathologization of spoiler fans. When dealing with cultural practices and cultural productions that are "not necessarily resistive but still less than normative" (Gray with Mittell, as cited in Gray, 2010, p. 149), and which may be written and spoken about via discourses of addiction, we should be extremely wary of moving too rapidly from the non-normative to the pathological as if the one can simply be inferred from the other. Given this, I am interested here in considering how spoilers can be theorized as part of everyday, routinized processes of anxiety and trust. To make this move, a brief detour through Anthony Giddens' use of the concept of "ontological security" will prove helpful (Giddens, 1990; 1991). Giddens argued that human subjects, if they are to avoid psychotic, pathological states, are required to develop a sense of trust in people and objects:

at the heart of the psychological development of trust, we rediscover the problematic of time-space distanciation.... Trust...brackets distance in time and space.... Trust, ontological security, and a feeling of the continuity of things and persons remain closely bound up with one another in the adult personality. (1990, p. 97)

Trust brackets distance, or "time-space distanciation" in Giddens' terminology, because once we've achieved an emotional state of trust in others we are able to tolerate their absence, relying upon their eventual return. As such, ontological security (and Giddens is drawing on Winnicottian psychoanalysis here) means the psychical attainment of basic trust in self-continuity and environmental continuity. Giddens further argued that ontological security is constantly (re)secured through social routines and habits:

Ontological security and routine are intimately connected.... When such routines are shattered—for whatever reason—anxieties come flooding in.... The continuity of the routines of daily life is achieved only through the constant vigilance of the parties involved—although it is almost always accomplished at the level of practical consciousness [i.e., is not fully conscious, though can be made so—MH]. (1990, p. 98)

In short, ontological security involves the reduction of anxiety, constituting "a continuing protective device (although fraught with possibilities of fracture and disjunction) against...anxieties" (Giddens, 1990, p. 99). Self-identity is one dimension of ontological security, and for Giddens this is narratively based:

A person's identity is not to be found in behaviour, nor—important though this is—in the reactions of others, but in the capacity *to keep a particular narrative going*. ...There is surely an unconscious aspect to this chronic "work", perhaps organised in a basic way through dreams.... A stable sense of self-identity presupposes the other elements of ontological security...but it is not directly derivable from them. (1991, p. 54; emphasis in original)

Roger Silverstone has related ontological security to television consumption, echoing Giddens' arguments about the importance of "the familiar and the predictable" (Silverstone, 1994, p. 19) and suggesting that TV as a technology, and a series of texts and genres, can help audiences (re)secure feelings of ontological security: "these attachments are over-determined...in television's case through its schedules, genres, and narratives. Its programmes are scheduled with consuming regularity" (Silverstone, 1994, p. 15).

For Silverstone, then, TV and its programs are "a contributor to our security" (1994, p. 19), meaning that "television is so much a fundamental part of our everyday life that it needs to be understood...both at a psychodynamic as well as at a sociological level" (1994, pp. 22–23). However, diametrically opposed to this security-securing reading is Robin Nelson's more recent take on what is considered "quality" TV such as emblematic HBO fare such as *The Sopranos*:

> The kinds of pleasure offered by formulaic television output offering "ontological security" differ markedly, however, from those at the other end of the spectrum where viewers might seek aesthetic visual pleasure or the frisson of risk. ...On subscription channels...there is no guarantee of limits...and thus, to those who seek it, there is a pleasure even in the anticipation of the transgression of social codes. Indeed, it might be claimed that they offer the pleasures of ontological insecurity. (2007, p. 19)

The major difficulty with this, and with Silverstone's counter-reading of ordinary TV's familiarity, is that each approach reads psychosocial dynamics of ontological (in)security as implied or dictated by the texts concerned. My suggestion is that fan spoilers are one audience practice and form of digital production that seeks to contain ontological insecurity and process it back into a sense of security. Along related lines, Rebecca Williams has recently theorized the endings of TV shows, suggesting that when a beloved fan object ceases broadcasting this may threaten fans' ontological security (2011, p. 274). But it is not only the "emotional void" (in Williams, 2011, p. 277) of a program's loss that may generate anxiety. Rather than binaries of ordinary/quality TV (or presence/absence of the fan object) being mapped onto ontological security/insecurity, it is important to consider the rhythms and experiences of always-on digital fandom. This is so because the knowledge that new official, canonical TV episodes are in production and are forthcoming, coupled with relatively little information about them, can also provoke fan anxiety. The idealized fan object is potentially threatened (in a way in which tie-ins, spin-offs, and unofficial material cannot pose a threat)—what if the new episodes are no good? What if unwanted story developments occur? Or much-loved characters leave? Rather than linking ontological insecurity only to transgressive TV drama (as Nelson has), or only to the end of a TV series (as Williams has), we should consider how psychosocial experiences of ontological insecurity can be woven into online fan practices.

Why might the forthcoming presence of a currently absent text pose such an issue for sections of fandom? Not because they are somehow neurotic or pathological, but rather because these fans' sense of self-identity—their "capacity to keep a particular narrative going" in Giddens' terms—is so firmly enmeshed with the narratives of their beloved TV show. Threats to diegetic narrative can thus be felt as threats to these fans' self-narratives: a process, which, as Giddens also observed, can have an unconscious aspect to it.

Spoilers offer one strategy, among others no doubt, for "weathering" threats to fans' self-narratives, which can emerge through potential or imagined threats to intensely cathected TV narratives (Giddens, 1991, p. 55). Indeed, the fan's very identity as a fan could be called into question by unwanted developments or characters in their show, meaning that forthcoming episodes will be freighted with both desire and anxiety (a balance that can, of course, be tipped further toward anxiety when major changes in production team or casting become known, e.g. a change of lead actor in *Doctor Who*, or *Torchwood*'s move to American co-production). Spoilers can provide a sort of "emotional inoculation" against fandom's anxieties (Giddens, 1990, p. 94). As Anthony Elliott has argued, "communication technology functions in part to deepen our reflexive scanning of unconscious anxieties" (1996, p. 152), and spoilers are produced through just such scanning. Elliott and Urry (2010) have more recently described how, via the use of smartphones, iPads and "miniaturised mobilities":

> the technological unconscious comes to the fore.... In classical Freudian theory, patterns of presence and absence primarily refer to significant others, such as parents, siblings, extended family and the like. With complex, network-driven systems, by contrast, we witness the emergence of various "virtual" others and objects resulting from the revolution of digital technologies. These virtual others and objects reconstitute the background to psychic experiences of presence and absence in novel ways. (p. 33; emphasis in original)

But patterns of presence and absence can now, via 24/7 digital fandom, relate to virtual objects such as to fans' favored texts, whose production can be intently and intensely monitored by online fan communities. Terming this only a "technological unconscious" profoundly misses the unconscious aspect that Giddens has linked to self-narratives. As such, spoiler fans would be better described as enacting a *technological-narrative unconscious*, using Web 2.0 to sustain ontological security, and crucially, to protect the story of their self-identity and *their self-narrative as a fan*. For fans who are processing ontological insecurity back into ontological security, spoilers are precisely about *not spoiling their relationship to the text*, but rather conserving and protecting their emotional attachments—guarding against disappointments, avoiding unpleasant shocks or surprises, and working-through possible threats to textual authenticity (and hence self-narrative).

Indeed, the very concept of a spoilerholic partly admits into consciousness the extent to which spoilers can act as mediators of anxiety and trust, and thus as "containing mechanisms...never free of the...anxieties and conflicts of the individuals that use them" (Elliott & Urry, 2010, p. 41). Again, though, it should be noted that this is not to pathologize spoiler fans: spoiler junkies cannot be assumed a priori to be addicts or neurotics. They are, rather, involved in re-securing feelings of self-continuity and object-continuity that are "closely bound up with one another in the adult personality" (Giddens, 1990, p. 97). This collision of ontological security and real-time narrative is evidently not restricted to spoiler fans; Daniel Miller has argued that Facebook's appeal to so-called "addicts" may be its provision of security, trust and reliability (2011, pp. 170–171) precisely linked to the "identification of Facebook time as narrative time" (2011, p. 192). Discourses of addiction may typically emerge at such points, where social media function psychosocially (rather than purely cognitively) in relation to a technological-narrative unconscious, with this process itself enabling self-narratives to be ordinarily grounded in ontological security.

Contra John Fiske's ethical opposition to psychoanalysis that I noted earlier, Sherry Turkle has recently suggested that "the ethic of psychoanalysis" may in fact be called for in relation to communications technology and social media: "Technology gives us more and more of what we think we want.... But if we pay attention to the...consequences of what we think we want, we may discover what we really want" (Turkle, 2011, pp. 284–285). Positing a psychodynamically split subjectivity, Turkle has reminded us that online practices can be somewhat conflicted. An example of this occurs for spoiler fans when they "feel the sting of regret, wishing that they hadn't known what was coming—although many suggest they cannot resist the temptation to seek out the spoilers for the next episode regardless" (Gray & Mittell, 2007), again indicating why discourses of the spoilerholic come into play. But such divided subjectivity is not a failure to attain pure rationality or empowering cognition (Jenkins, 2006); it can be read as a desire for ontological security—protecting the fans' self-narrative and investment in a series' diegesis—which trumps the "normative plot-centric" pleasures of TV consumption for some fans (Gray with Mittell, as cited in Gray, 2010, p. 152). And, to be sure, viewing TV in specific ways remains culturally normative. As one TV showrunner has put it, symbolically attacking spoiler fans:

> And hello the internet forums. ...some ghastly little show-off, who was lucky enough to be at our press night, has typed up the entire plot of Episodes 1 and 2 in the most bungling, ham-fisted English you can imagine, and put it where everyone can see it. ...All our months of work and effort, flattened and smeared and smudged into a dreary page of badly-chosen words.... Stories are better if you don't know what's coming. More exciting, more real. (Moffat, 2011a, p. 6)

Such stories may perhaps be "more real," but they nevertheless run the anxiety-provoking risk of rendering fans' self-narratives less real by calling them into question and by generating ontological insecurity. What if the Steven Moffat era of *Doctor Who* undoes all the good work of the show's 2005 relaunch? What if a new series (series six being referred to here) alters *Doctor Who*'s textual identity and casts doubt over its authenticity and success? It is notable that these spoilers concern a moment of transition in the show— the beginning of a new run of episodes.

If they're not reinforcing discourses of addiction, then attacks on spoiler fans seem instead to allege pathological narcissism—"a ghastly little show-off" in Steven Moffat's ghastly little expression. The problem that's raised is thus a social one: fans won't keep spoilers to themselves, but insist on sharing them online, presumably to gain fan cultural status. In a similar discursive mode, *Doctor Who*'s current producer, Marcus Wilson, has said: "This year, some fans revealed dialogue from Episode 13, which was depressing. ...I'd rather they just knew themselves, rather than telling everyone" (in Cook, 2011, p. 22). But as well as forming a knowledge community or collective intelligence, as Jenkins (2006) argued, spoiler fans also gain "the *feeling that they are needed*" (Baym, 2010, p. 86; emphasis in original). As Nancy Baym has pointed out:

> On the board for my favorite band, one member gained high standing because he regularly searched the internet for relevant videos and photographs and then shared them with the others, solving the ongoing informational problem fans face of never being able to know enough about that which they love. ...People assume roles by enacting consistent and systematic behaviors that serve a particular function. (2010, p. 86)

The ongoing fan informational problem of processing ontological insecurity into ontological security can thus involve a further dimension of emotional security—that of adopting a role within the fan community and so feeling needed. In a sense, the (re)securing of ontological security is over-determined in this instance, being produced through the very routines of spoiler-discovery, debate and fan cultural production as well as through the content of specific spoilers. David Gauntlett noted in his definition of creativity that it "is something that is *felt*" (2011, p. 79; emphasis in original), and I would say that the same is true of fan spoilers, rather than these acting as a purely factual or informational exercise. The emotions, anxieties, conflicts and highs of spoiler pursuit hence seem rather underplayed and under-theorized by models of collective intelligence (Jenkins, 2006).

According to Gauntlett, "While 'sharing' and 'connecting' themes sound both warm and benign...seeking recognition...includes a harder edge, a kind of demand—'notice me!'—which we have to understand as well" (2011, p. 101). Spoilers can have this kind of "notice me!" aspect, but then so too does being a television showrunner and writing "Production Notes" for *Doctor*

Who Magazine; fans hardly have a monopoly on showing off. Rather than glibly equating fans and producers in this negative sense, though, I think the contrast that Steven Moffat drew between the official text and unofficial spoilers is rather more instructive. The TV series should supposedly be "exciting...real," whereas spoilers are "dreary...badly-chosen words." What's at stake here is more than an attack on fan narcissism or improper demands for recognition: the TV showrunner is defending his properly tutored, skillful expert creativity against fans' unskilled, illiterate and improper amateurism. What's at stake is precisely what David Gauntlett sought to dematerialize in *Making Is Connecting*, via his definition of creativity as "not something that needs external expert verification" (Gauntlett, 2011, p. 79)—that is, a concept of creativity as requiring skill and expertise.

In Moffat's production discourse, spoilers are allegedly mere reproductions; feeble shadows of the text's professional creativity and aesthetics. They spoil because they lack skill, and craft, and good judgment. In contrast, professionals know just how to tease the audience with trailers and snippets of information, e.g., a pre-broadcast promise that "One of these four characters [Amy Pond, the Doctor, River Song, Rory Williams] will die in the first episode of the new series!" (a fake spoiler, as it eventually transpired; Spilsbury, 2011, p. 5). Or spoilers are brute facts, lacking the proper narrative and aesthetic context:

> We don't want photos of the monsters in the papers before they're seen on screen...[as a result of journalists trawling fan forums]. To see a monster on set, having a cup of tea, not lit properly, destroys the impact of seeing it for the first time on TV, in all its glory. (Wilson in Cook, 2011, p. 22)

Again, fan spoilers appear to lack tutored, skillful creativity. Unlike other forms of fan cultural production, spoilers thus seem distanced from logics of creativity—they may be about speculation, about collective intelligence or interactivity (Booth, 2010; Jenkins, 2006), but they haven't yet been strongly theorized as a form of creativity, just as they are ideologically denied this status in production discourse (which defends commodity texts and brands against incursion). Even given Gauntlett's generous account of digital creativity, it is unclear that online spoilers would make the grade:

> making and sharing our own media culture—I mean via lo-fi YouTube videos, eccentric blogs, and homemade websites, rather than by having to take over the traditional media of television stations...isn't quite "by hand" in the literal pottery-making, woodcarving sense, but...I feel is still basically a handicraft which connects us with others through its characterful, personality-imprinted, individual nature. (Gauntlett, 2011, p. 18)

As collective intelligence, spoiler production significantly seems to lack this sense of handicraft or eccentric, personality-imprinted production, After all, it is the acquisition of information about a TV series; hardly a bohemian activity. Against all such assumptions, I want to conclude by suggesting that rather than being narcissistic or a derivative copying, spoiler fans' practices can be viewed as highly creative. The "notice me!" aspect of such fan practices can thus be theorized not as a negative attribute, but rather as the result of skillful creativity (akin to that of official media production, in fact). Spoiler fans who gain recognition in fan circles (e.g., on forums, blogs and Twitter) do so not only for being accurate, reliable and trusted sources—they do so also as a result of getting the photo that was difficult to take, and so capturing a moment on location; obtaining information via skills of persistence, audacity, and a feel for the game; witnessing filming via dedication and hard work; and collating and presenting information with a "flair for design" (Gauntlett, 2011, p. 232). Such activities may well be adversarial, as in the case of what was called "Rivergate"—*Doctor Who* fans revealed the identity of River Song before series six of the program had even begun broadcasting—but they are nevertheless far more then mere narcissism, and more than normotic fact-collecting.

In this chapter, I have argued for the importance of adopting a broadly psychoanalytic or psychosocial approach to spoilers. Spoilers are more-or-less pathologized in production discourse (Moffat, 2011a; 2011b), and their non-normative status may lead us to radically misread fan discourses of the spoilerholic. But it remains perfectly possible to interpret fans' routines of spoiler production and consumption outside these pathologizing discourses, while still recognizing the emotions, anxieties, conflicts and creativities of this dimension of digital fandom. Especially through the conversion of ontological insecurity into feelings of security, and the containing of anxiety, spoilers can have a value psychosocially as well as communally. And through their relationship to temporality—a sort of generalized just-in-time or always-on digital fandom—as well as through their exercising of skilled "everyday creativity" (Gauntlett, 2011, p. 76), spoilers might come to seem less marginal, less strange, and a more ordinary, routinized part of many online fan cultures. Finally, I have suggested that for fans who embrace the world of spoilers, this can be a way of conserving their fan identities, and so unconsciously working to protect fan self-narratives themselves from being spoiled.

References

Bailey, S. (2005). *Media audiences and identity: Self-construction in the fan experience.* New York, NY: Palgrave Macmillan.

Bainbridge, C., & Yates, C. (2010). On not being a fan: Masculine identity, DVD culture and the accidental collector. *Wide Screen, 1*(2). Retrieved February 27, 2012, from http://widescreenjournal.org/index.php/journal/article/view/39/80

Baym, N. (2010). *Personal connections in the digital age.* Cambridge, England: Polity Press.

Black, R. W. (2008). *Adolescents and online fan fiction.* New York, NY: Peter Lang.

Bollas, C. (1987). *The shadow of the object: Psychoanalysis of the unthought known.* London, England: Free Association Book.

Booth, P. (2010). *Digital fandom: New media studies.* New York, NY: Peter Lang.

Bury, R. (2005). *Cyberspaces of their own: Female fandoms online.* New York, NY: Peter Lang.

Cook, B. (2011). Baptism of fire! *Doctor Who Magazine, 441,* 18–22.

Craib, I. (1998). *Experiencing identity.* London, England: Sage.

Cromarty, D., & Lewin, R. (2011). Are you a spoilerphobe or a spoilerholic? *SFX Bookazine 2: Doctor Who: The Fanzine, 68.*

Elliott, A. (1996). Subject to ourselves: Social theory, psychoanalysis and postmodernity. Cambridge, England: Polity Press.

Elliott, A., & Urry, J. (2010). *Mobile Lives.* New York, NY: Routledge.

Fiske, J. (1990). Ethnosemiotics: Some personal and theoretical reflections. *Cultural Studies, 4*(1), 85–99.

Freud, S. (1901/1976) *The psychopathology of everyday life.* London, England: Penguin.

Gauntlett, D. (2011). *Making is connecting: The social meaning of creativity, from DIY and knitting to YouTube and Web 2.0.* Cambridge, England: Polity Press.

Giddens, A. (1990). *The consequences of modernity.* Cambridge, England: Polity Press.

Giddens, A. (1991). *Modernity and self-identity: Self and society in the late modern age.* Cambridge, England: Polity Press.

Gray, J. (2010). *Show sold separately: Promos, spoilers and other media paratexts.* New York, NY: New York University Press.

Gray, J., & Mittell, J. (2007). Speculation on spoilers: Lost fandom, narrative consumption and rethinking textuality. *Participations: Journal of Audience & Reception Studies, 4*(1). Retrieved March 4, 2012, from http://www.participations.org/Volume%204/Issue%201/4_01_graymittel l.htm

Gray, J., Sandvoss, C., & Harrington, C. L. (2007). Introduction: Why study fans. In J. Gray, C. Sandvoss, & C. L. Harrington (Eds.), *Fandom: Identities and communities in a mediated world* (pp. 1–16). New York, NY: New York University Press.

Harrington, C. L., & Bielby, D. D. (1995). *Soap fans: Pursuing pleasure and making meaning in everyday life.* Philadelphia, PA: Temple University Press.

Hills, M. (2002). *Fan cultures.* London, England: Routledge.

Hills, M. (2010). *Triumph of a Time Lord: Regenerating Doctor Who in the twenty-first century.* London, England: I. B. Tauris.

Hoxter, J. (2000). Taking possession: Cult learning in *The Exorcist.* In X. Mendik & G. Harper (Eds.), *Unruly pleasures: The cult film and its critics* (pp. 171–185). Guildford, England: FAB Press.

Jagodzinski, J. (2004). *Youth fantasies: The perverse landscape of the media.* New York, NY: Palgrave Macmillan.

Jagodzinski, J. (2008). *Television and youth culture: Televised paranoia.* New York, NY: Palgrave Macmillan.

Jenkins, H. (1992). *Textual poachers: Television fans and participatory culture.* New York, NY: Routledge.

Jenkins, H. (2006). *Convergence culture: Where old and new media collide.* New York, NY: New York University Press.

Miller, D. (2011). *Tales from Facebook.* Cambridge, England: Polity Press.

Minsky, R. (1998). *Psychoanalysis and culture: Contemporary states of mind.* Cambridge, England: Polity Press.

Moffat, S. (2011a). Production notes. *Doctor Who Magazine, 434,* 6.

Moffat, S. (2011b) *Doctor Who boss 'hates' fans who spoil show's secrets.* Retrieved March 3, 2012, from http://www.bbc.co.uk/news/ entertainment-arts-13353367

Nelson, R. (2007). *State of play: Contemporary "high-end" TV drama.* Manchester, England: Manchester University Press.

Odden, K. (1998). Retrieving childhood fantasies: A psychoanalytic look at why we (re)read popular literature. In D. Galef (Ed.), *Second thoughts: A focus on rereading* (pp. 126–151). Detroit, MI: Wayne State University Press.

Samuels, R. (2010). *New media, cultural studies, and critical theory after postmodernism.* New York, NY: Palgrave Macmillan.

Sandvoss, C. (2005). *Fans: The mirror of consumption.* Cambridge, England: Polity Press.

Schäfer, M. T. (2011). *Bastard culture! How user participation transforms cultural production.* Amsterdam, The Netherlands: Amsterdam University Press.

Silverstone, R. (1994). *Television and everyday life.* London, England: Routledge.

Smith, J. (2010). *Withnail and us: Cult films and film cults in British cinema.* London, England: I. B. Tauris.

Spilsbury, T. (2011). Letter from the editor. *Doctor Who Magazine, 433,* 5.

Tryon, C. (2009). *Reinventing cinema: Movies in the age of media convergence.* Piscataway, NJ: Rutgers University Press.

Turkle, S. (2011). *Alone together: Why we expect more from technology and less from each other.* New York, NY: Basic Books.

Williams, R. (2011). "This is the night TV died": Television post-object fandom and the demise of *The West Wing. Popular Communication, 9*(4), 266–279.

CHAPTER EIGHT

Seven Stories from the "It Gets Better" Project: Progress Narratives, Politics of Affect, and the Question of Queer World-Making

Gust A. Yep,[1] Miranda Olzman, and Allen Conkle

In the era of social media—in a world with YouTube and Twitter and Facebook—I [can] speak with [lesbian, gay, bisexual, transgender and queer] kids right now. I [don't] need permission from parents or an invitation from a school. I [can] look into a camera, share my story, and let [lesbian, gay, bisexual, transgender, and queer; henceforth, LGBTQ] kids know that it got better for me and it would get better for them too. (Savage, 2011, p. 4)

Creating user-generated content (UGC) to reach mass audiences through the powerful intersection of production and consumption of new media and responding to the ongoing symbolic and physical violence inflicted upon gender non-conforming and sexual minority youth, syndicated columnist and author Dan Savage and his partner Terry Miller decided, in September 2010, to post a video on YouTube to offer hope and encouragement to this group, marking the beginning of the It Gets Better (IGB) Project. As Savage and Miller's video went viral, numerous individuals and groups, including President Barack Obama, Secretary Hillary Rodham Clinton, Congresswoman Nancy Pelosi, Bishop Gene Robinson, gender activist Kate Bornstein, British Prime Minister David Cameron, and members of the Broadway and New York Theatre Company, among many others, started posting their own messages of hope for gender and sexual minority youth. By June 2011, over 10,000 user-generated videos had been posted and viewed over 35 million times (http://www.itgetsbetter.org/

pages/about-it-gets-better-project/). The unprecedented response from the worldwide audience has prompted the creation of the IGB Project Web site complete with information about events, ways of getting help, activist opportunities, and IGB-related merchandise. In March 2011, *It Gets Better: Coming Out, Overcoming Bullying, and Creating a Life Worth Living,* edited by Savage and Miller, was published by Dutton.

A number of critics (e.g., Halberstam, 2010; Puar, 2010; 2012), however, emphatically contest the message—"things get better"—of the IGB project. For example, Halberstam (2010) noted that things get better for mostly privileged white gay men, who represent mainstream gay voices. For queer people of color, trans-identified individuals, queers living with disabilities, working class sexual minorities, genderqueers (people whose self-presentation challenges the gender binary system), and gender non-conforming youth—that is, the non-mainstream voices of LGBTQ communities—things do not necessarily get better (Halberstam, 2010; Puar, 2010). Using queer theory as a conceptual and methodological framework, this chapter examines seven audience-generated stories from the IGB Project, posted by non-mainstream individuals in LGBTQ communities in the United States. More specifically, our analysis focuses on the nature of progress narratives, the politics of affect, and the question of queer world-making in these stories. After an overview of our conceptual and methodological framework derived from queer theory, we discuss discourses of progress in the IGB Project before turning to examine how progress is characterized in the seven stories, consider the affective investments of the authors, discuss the notion of "queer world-making" as postulated by queer theorists, and explore the potentials and pitfalls of audience-generated media aimed at offering hope and support for gender non-conforming and sexual minority youth.

Queer Theory and Queer Methodology

Influenced by poststructuralism, deconstructionism, and postmodernism, queer theory emerged as a significant approach for engaging in social action, participating in political activism, and contributing to cultural politics (Yep, 2009). Using a set of tools to examine and understand social relationships, particularly those organized around constructions of sexuality and desire that highlight the centrality of power in historical and geopolitical contexts, queer theory is a particularly appropriate conceptual framework for our analysis of the stories posted in the IGB Project.

Since its inception, work in queer theory can be characterized in terms of generations. Although not meant to suggest a linear and unidirectional development of ideas but a mutually influencing exchange and an ongoing critique and response to each other, "first-generation" (Yep, 2009, p. 818) research generally focuses on the analysis of heteronormativity in social relations. According to Yep (2003), heteronormativity refers to the structures of understanding, relational assumptions, cultural discourses, and social institutions that construct heterosexuality as privileged, morally right, taken-for-granted, coherent, and stable. In short, heteronormativity is the normalization of heterosexuality as an institution, identity, practice and experience. "Second-generation" work in queer theory (Yep, 2009, p. 818) continues to interrogate heteronormativity, but broadens the analytical domain to examine multiple forms of normativity—such as racial, class, gender, erotic, and national assumptions that produce urban and body normativities and U.S. homonationalism, among others. Such research attends to how race, class, gender, sexuality, ability, and nation intersect and mutually constitute each other to produce specific types of subjects and subjectivities. Homonormativity, for example, has become an important concept for the analysis of gay and lesbian lives, experiences, and politics in an era of neoliberal global capitalism (Yep & Elia, in press). According to Duggan (2003), homonormativity refers to the production and normalization of de-politicized gay and lesbian identities based on domesticity and consumption and the reification of heteronormative practices and institutions. The celebration of individual freedom and individual responsibility as opposed to collective action, community building, and support of the public good; the celebration of assimilation to heteronormative ideals as opposed to contestation of their violence on gender nonconforming people and sexual minorities; and the celebration of visibility through consumption and the creation of gay and lesbian niche markets as opposed to visibility through political change in major societal institutions are the hallmarks of what Duggan (p. 50) has coined "the new homonormativity." For our analysis, we use concepts –progress narratives, politics of affect, and queer world-making—from both "first-generation" and "second-generation" queer theory (Yep, 2009, p. 818).

Research in queer theory, according to Browne and Nash (2010), does not lend itself to a singular and definitive "queer method." Our attempt to understand stories about queer lives, as represented by audience-generated IGB Project videos, uses what Judith Halberstam (1998, p. 13) called "a scavenger methodology," which seeks to "collect and produce information on subjects who have been deliberately or accidentally excluded [such as non-mainstream voices in LGBTQ communities] from traditional studies of human behavior" (Halberstam, 1998, p. 13).

We selected seven stories provided by the video clips uploaded to the IGB Project Web site, representing an intricate mix of sexuality intersecting with race, class, gender, age, (dis)ability, and religion within the U.S. geopolitical context in 2010–11. Taken together, they represent a rich tapestry of diverse social locations, life experiences, individual subjectivities, personal journeys, individual life paths, and personal and collective visions; in short, they provide a glimpse of a queer life—at this historical juncture and in its imagined future. The videos and their creators are:

AbleTraveler, a self-identified white gay male in his 50s, physically disabled, chronicles the complexities of growing up in a small U.S. rural town and discusses dealing with the interlocking oppressions that are inherent in his intersecting identities. He offers a complex progress narrative.

AJ541 is one of four African American gay men discussing their lives, which include a suicide attempt, experiences with getting bullied, feelings of difference, and ways of finding support. They offer hope and illustrate the concept of chosen families. Their progress narrative complicates the IGB Project by placing impetus on the viewer to find ways to create a better life instead of waiting for it to get better.

blackgayandjewish, a lesbian-identified Black Jew in her mid 30s, discusses the complications of growing up as a Black Christian woman who came into her sexuality in her late 20s. She explores the problems inherent in a progress narrative and the complications inherent in the intersections of identity that complicate material human experiences.

dudessforrent, a Chinese American lesbian who self-identifies as genderqueer, discusses her history with bullying and the acceptance and normalization of violence in her life. She offers words of hope for youth who are contemplating suicide.

lecercle01, a gay-identified Shiite Muslim Pakistani man in his 20s, talks about his responsibility to speak out as a generally unheard voice in the U.S. He discusses his complicated relationship with stereotypical gay gender performances as well as the negotiation of being gay, as a part of familial, cultural and religious identities.

TheKairlelise, a white lesbian transwoman, discusses her journey which includes suicidal thoughts, transition from a biological male to a self-identified woman, and becoming a lesbian when she found "the woman of her dreams." She offers both her personal and public information for youth to find support.

victoriaray, a self-identified lesbian of color perceived as Latina, shares the struggles with growing up poor and queer. She chronicles the physical and emotional violences she experienced when her mother caught her kissing a girl. Her mother attempted to "beat the gay out of her" and at 15, she was kicked out of her home and forced to survive on the streets.

Narratives of Progress and the Promise of Happiness

The notion that things get better suggests a powerful utopian narrative of progress about the history of LGBTQ people in the social world. Such a narrative maintains that lesbians and gays—a group that has been socially and culturally categorized as deviant—began to organize, build community, talk back, and demand acknowledgment and legitimacy by the larger heteronormative culture, in what Michel Foucault (1978) called "the formation of a 'reverse' discourse" (p. 101). This discourse exemplifies the reversibility of power in which a dominated and stigmatized group turns its own subjugation and shame into affirmation and pride. Contemporary lesbian and gay history and identity are informed by this reverse discourse with progress as its corollary. Progress is imagined as a boundless and irresistible advancement of humankind (Benjamin, 1968). A narrative of progress asserts that lesbian and gay life will continue to improve. This narrative is so powerful, as Heather Love (2007, p. 3) accurately has pointed out, even though "many queer [theorists and] critics take exception to the idea of a linear, triumphalist view of history, [they] are in practice deeply committed to the notion of progress," concluding that "despite [their] reservations, [they] just cannot stop dreaming of a better life for queer people." Implicit in this narrative is "the promise of happiness," to invoke Sara Ahmed's (2010) words, which suggests that happiness will follow "if you have this or that, or if you do this or that" (p. 29).

Although the seven stories deploy narratives of progress in a variety of ways, they center around four themes. The first three propose ways to achieve progress—how it gets better—and the last questions the potential for such achievement—whether it gets better. Together, the four themes are: (a) the rhetoric of acceptance that often suggests—either implicitly or explicitly—that biological families and higher spiritual powers will embrace the young LGBTQ person; (b) the idea of a chosen family as a path—either directly or indirectly—to making oneself happy; (c) the concept of success—personal and/or professional—that promises to bring happiness to the individual; and (d) the possibility of a better life—individually and/or collectively—in the future. These themes seem to suggest that finding acceptance, developing a relational network, and achieving success—major features of a narrative of progress—will, in many ways, bring happiness to the life of the LGBTQ person. We present our themes as questions to indicate that there are variations and, at times, considerable disagreement, between the seven stories; and, consistent with queer theory, to suggest their somewhat uncertain, provisional, and fluid nature.

Acceptance?

Based on our reading of the seven stories, acceptance can be defined as the process of developing and sustaining a positive regard for one's experiences (e.g., same-sex attraction), identities (e.g., sexual, racial, gender, class, religious, etc., as a complex mixture), and behaviors (e.g., gender performances). The stories tend to focus mostly on acceptance at the individual (self-acceptance and coming to terms with one's sexuality and gender performance) and spiritual (acceptance by God or a higher power) levels, often neglecting questions of acceptance at the social, cultural, and structural levels.

Individual level acceptance places much of the focus and responsibility on the LGBTQ person. For example, one of the four African American gay men in the video posted by AJ541, reminds us, "just remember that you are the captain of your own soul, that you are the one that determines what happens to you, it's you that determines your fate." Viewers are thus to take charge of their destiny by creating a future of their own design. To do this, blackgayandjewish suggests that the most important thing—indeed, "the only thing that matters," as she put it—"is that you are true to yourself, that you love yourself...." Visually suggesting that self-reliance and perseverance pay off in the construction of one's own future, TheKairielise's video begins in black and white—when she narrates her struggles as a non-normative cis-gendered male—and transitions to color when she recounts the completion of her journey to become a lesbian transwoman.

In addition to focusing on the individual, the stories also point to acceptance at the spiritual level. Not surprisingly, images of an all-accepting God are commonly brought up, sometimes in numerous ways within the same posting. For example, blackgayandjewish reminds the viewer, more than once, that "God loves you" and goes on to describe her experiences in a Catholic high school: "I have a problem with some people in the Catholic Church, but I don't have a problem with the Catholic Church." Another video, posted by lecercle01, challenges the automatic religious condemnation of homosexuality when he states "[d]on't worry if you're just really, ah, conflicted about your religious beliefs, I don't think your sexuality is necessarily incompatible with, ah, Islamic teaching specifically" before he offers his own understanding and interpretation of Islamic Syrian law. In yet another posting, victoriaray, asserts that, "God is perfect. And he created all of us, and he created all of us as perfect as he intended, so don't let anyone tell you that you're different."

Chosen Family?

A theme associated with a narrative of progress suggests that LGBTQ people now have the option of creating their own families—a family of choice—when rejected by biological families. A chosen family is a kinship network that could include friends, romantic partners, coworkers, or others.

Some chosen families might be created by carefully selecting people who support one's identities and experiences. AbleTraveler, using his own experiences to express empathy to a presumably LGBTQ viewer, observes,

> For those that have a disability, you know that you are already different. Feeling separate, unwanted, and not belonging only adds to your pain. After becoming aware that you are also gay doesn't help. Living in a small town where you believe that you are the only gay and disabled person on the planet doesn't help your self-esteem.

To offer encouragement and support, AbleTraveler provides some suggestions:

> There are tons of connections and resources out there to help you. Tap into the internet search engine, and support...you know I support a person with disability that happen to be gay, lesbian, bi, trans, queer and questioning. We support each other in many ways to create intentional family of those that love and care for us.

The intentional creation of a loving network of support—a chosen family—offers hope that life will get better in the future.

Sometimes chosen families emerge inadvertently. After her mother saw her kissing a girl and forced her to leave home, victoriaray was living on the streets when she found "a couple of awesome people like my friend Lupe, my aunt Judy, that would always have a special place in my heart, that took care of me."

Success = Happiness?

Based on our reading of the seven stories, success seems largely defined in normative terms—that is, personal and professional achievements valued by heteronormative society. Exemplifying this concept of success, TheKairielise, a white transwoman, said:

> I moved out on my own, got a full time job, started making my own money and, in 2007, I started living as the woman I always should have been . . . I fell in love and now I completed my journey.

This narrative shows her integration as a "non-normatively sexed/gendered [person] into heternormative capitalist society" (Irving, 2008, p. 51) by focusing on her economic productivity, physical transformation into a more gender-normative woman, and romantic coupling. It appears that adopting normative values—making money, becoming independent, finding love—produces a successful and happy subject in spite of her differences as a gender and sexual minority.

In some stories, success is rooted in professional accomplishments. Although the four African American gay men in a video posted by AJ541 note that "we don't want to dictate, you know, how our lives are because of the material things," they highlight economic productivity while downplaying sexual non-normativity:

> We do have very successful things going on. I'm a director of finance, David, uh, to my right does PR, we have a journalist behind us, and you know [the fourth man in the video is a] business manager . . . Um, so we're very successful in that aspect but that's not what makes us successful but what makes us successful, you know, is that our gayness doesn't define who we are, it's a part of us but it doesn't define who we are.

Success, as exemplified by professional accomplishments, has been linked to "things getting better" and the possibility of a happy life. For example, AbleTraveler says:

> As a gay and disabled man that graduated from high school and went on to college and then developed a career, working in banking, the Boeing company as Industrial Engineer, hired as a program support specialist, the Federal Court as the PC systems administrator, and then later the Department of Education as the Customer Service Manager, I can tell you that it does get better.

He hints at the promise of happiness when he indicates that the viewer "will be valued by others" and tells people to "choose hope and happiness that is part of your future." This promise, Ahmed (2010, p. 32) notes, gets individuals to follow a socially prescribed trajectory, such as becoming personally and professionally successful, "where the 'there' [e.g., feeling valued by others] acquires its value by not being 'here' [e.g., feeling bullied by others]."

Does It Get Better?

Although the seven stories seem to present a narrative of progress through notions of acceptance, chosen family, and success, they do not collectively and unequivocally offer the promise that life will get better. However, AbleTraveler seems to capture a common sentiment in the videos we examine when he points out that "nothing is perfect, but . . . but so much progress has been made."

What does this progress look like? Do things get better for gender non-conforming and sexual minority youth? Some stories assert that recent historical changes have made life easier for this group; in short, they enthusiastically claim that things do get better. These positive proclamations seem to make sense in the context of the IGB Project—individuals and groups are explicitly and implicitly invited to share their messages of hope and optimism for public consumption. Therefore, it is not surprising that this is the general tone of a number of the stories we examine. For example, one of the African American gay men, in a video posted by AJ541, observed:

> [T]he world is changing too.... we have shows that are gay. I came out when I was fourteen when no one was gay, when Ellen came out on television it was not the cool thing to do . . . [now] there are people out there that can help you and can be there for you and that can support you and that can love you for who you are. So it does get better, it really does.

For this group of men, life "gets way better, extremely better, all the time better...." Similarly, TheKairielise says that:

> [t]here are so many [people] out there that will listen and they will help you get through this. I know you are valuable to the world, I know you have so much to contribute. I made it, so can you. It does get better, I promise.

These stories offer encouragement and hope to the presumably LGBTQ person through a narrative of progress about a changing and more accepting social world manifested through the increasing number of caring and supportive people.

Other stories, however, are much more cautious about promises of a better life. In his posting, lecercle01 introduces himself in the following manner:

> I grew up in a Syria Muslim Pakistani household, which is not the easiest circumstances to grow up gay and I seemed to be the only gay Muslim I know of, but I know that with 1.5 billion people in the world, ah, I can't possibly be the only one, so if you are out there, you are going through a really hard time, I am here. I exist, people like you exist; you are not alone.

After sharing his struggles with his family and religion, he ends his video not with a promise of a better life but with an assurance that the viewer is not alone. He says,

> Don't be afraid of your parents, family, umm, I would hope . . . that they will respond kindly. I can't promise. I am of a specific situation but I think it all worked out for the best, I am in a really good place now. I am two blocks away from the Hudson River Park where I go on a sunny day, and I see gay couples of all shapes and sizes and colors, with their children . . . [it] really shows a hopeful future for our country, and people like us specifically and what we can grow up to be, so I plea with you, I [ask] you to hang in there, and umm, and to let you know that you're just not alone.

In another story, blackgayandjewish is much more explicit about not offering hopeful promises,

> My main message to you is, I am not going to tell you it gets better and everything is rosy after high school because it is not . . . [t]here [are] still dumb people out there, really hateful people out there . . .

In the above narrative, life does not necessarily and inevitably get *better* for gender non-conforming and sexual minority youth. It simply gets *different*.

Feeling Backward

Many lesbian and gay people, including scholars and activists, refuse to engage with the traumatic violence of the past. Exemplifying this refusal, when Dan Savage (2011) conceived the IGB Project, he asked: "What [is] to be gained by looking backward? Why dwell on the past?" (p. 3). Although the past cannot be changed and the bullying could not be stopped, the refusal to look backward, according to Heather Love (2007), "may entail other kinds of losses" (p. 11), and "makes it harder to see the persistence of the past in the present" (p. 19). In other words, without engaging past degradation and trauma, lesbian and gay people will continue to be haunted by the past. Arguing for the need to address the negative affect associated with the trauma of violence—symbolic, psychological, and material—we believe that as we move forward, we must also look back, similar to Benjamin's (1968) angel who is propelled into the future while glancing backward as "the pile of debris before him grows skyward" (p. 260), to examine the "archive of feelings" (Cvetkovich, 2003, p. 7) resulting from trauma. To put it another way, we need to engage in "feeling backward," to invoke Love's (2007) words, and we do so by examining the affective archive highlighted and downplayed by the stories. Feeling backward, in this context, involves naming and working through the affect that is part of queer life in a homophobic and heteronormative world. Our reading of the stories yields a continuum of affect ranging from negative to positive.

Ugly Feelings?

Originally used by Sianne Ngai (2007) to describe seemingly minor and politically ambiguous feelings (such as envy, anxiety, and irritation, which are perceived as potentially blocking individuals and groups from doing things) that are distinguished from useful feelings (such as anger and fear, which are seen as motivating individuals and groups to engage in action), we expand the construct to include a wide range of negative affect: anger, fear, rage, shame, despair, withdrawal, loneliness, resentment, self-hatred, envy, anxiety, irritation, and others. Ugly feelings, in this context, refer to a range of affect considered socially and culturally negative, regardless of their political utility.

Gender non-conforming and sexual minority people experience continuing and unrelenting violence in a homophobic and heteronormative world (Yep, 2003). Such violence can be symbolic (e.g., a genderqueer individual who does not fit into the gender binary), psychological (e.g., bullying), material (e.g., physical abuse), or, frequently, a combination thereof. It is hardly surprising that all seven stories are, in many ways, about living and enduring these multiple forms of violence but it is astonishing how the narrators discuss such experiences. Often, violence is normalized and ugly feelings downplayed.

Dudessforrent describes the pervasive symbolic and material violence in her life in the story of how fellow students hit her head with a rock, at the age of six, because she was "different." She describes ongoing bullying at school, getting "beat up" and "punched," and harassed by the police for going into a woman's bathroom, and confesses to the viewer, "I never thought that I would live this long. I thought I would never live past twenty-five years old. Um, I thought I would be dead, for sure." After the confession, she expresses surprise that other people were upset about the violence:

> They told the dean, the dean called the police who asked me to do a formal report so they could start an investigation and that's when I realized that even though I didn't know any of these people that they were very, very, very supportive of me. I didn't think anybody, anybody would care but I was wrong.

Her surprise that strangers would be angry about the ongoing violence—a major theme of her story—suggests how violence was normalized and expected in her daily existence. She mentions that she was "upset" and "really scared" but downplays these feelings and highlights her surprise that so many presumably straight strangers and coworkers cared. The tendency to steer away from trauma, or a wound inflicted upon the mind, has potential consequences, as Caruth (1996) has noted:

> [T]rauma is not locatable in the simple violent or original event [or series of events] in an individual's past, but rather in the way that its very unassimilated nature—the way it was precisely not known in the first instance—returns to haunt the survivor later on. (p. 4)

This haunting might be one of the losses from refusing to engage with the difficult and painful queer past (Love, 2007).

This refusal pervades other stories as well. Victoriaray shares her story of abuse and violence—this time, the tormentor was her mother. After her mother witnesses her kissing another girl, victoriaray explains,

> When I got home that night, ah, [my mother] was waiting for me, kinda disgusted that she saw her first-born kissing a girl, she decided to beat the gay out of me. And she decided to do that by beating me with a stick and when the stick didn't beat the gay out of me, she beat me with a hammer. And when the hammer didn't beat the gay out of me, she beat me with her hands. And she was unsuccessful then, she asked me to leave the house. So, at 15 years old, I was out on the streets.

In recounting her story, the tensions of feeling backward seem evident as victoriaray struggles to examine her pain (choking up at various moments and holding back her tears) and attempts to leave it behind (speeding up when explaining a difficult part of her story and slowing down to talk about the people who helped her through).

Although the refusal to linger in the archive of feelings of the past is certainly understandable—who wants to re-live the bullying, humiliation, and abuse?—it creates a paradox for the abused. Steering away from the abusive past sustains the very conditions that keep it troubling, or in Caruth's words (1996), "the story of a wound that cries out" (p. 4) with an "endless impact on a life" (p. 7).

Cruel Optimism?

According to Lauren Berlant (2011), optimism is a force that moves people out of themselves and into the social world in order to get closer to something they cannot accomplish on their own but sense and feel in the wake of a person (e.g., someone telling a LGBTQ individual that things got better for him or her) or a way of life (e.g., a bullied and abused LGBTQ person watching a video of a lesbian or gay couple living a happy and fulfilling life). Optimism becomes cruel when the very thing desired (e.g., to openly love a person of the same gender) is itself an obstacle to one's flourishing (e.g., living in a social world that continues to pathologize and discriminate same-gender love relations).

In some ways, the IGB Project videos may potentially produce a relation of cruel optimism with their audience. For example, to promise presumably LGBTQ young viewers that "it does get better" (e.g., AbleTraveler, AJ541, TheKairielise, victoriaray), without knowing the complex particularities of their lives—the nature and severity of the abuse and torment they might be experiencing, the social networks they may or may not have, the material conditions of their lives, their life chances based on race, class, gender, sexuality, and (dis)ability hierarchies in our culture, among many other factors—is, in many ways, an empty promise. That such a promise might help young LGBTQ people get through a challenging period of their lives only to find themselves facing more challenges in the future, such as social prejudice, cultural stigma, individual and structural discrimination, among others, can be misleading, illusory, and potentially cruel.

Although we do not suggest that messages of hope and optimism, particularly in times of struggle, should be avoided, we encourage deeper reflection into the motives, hopes, and aspirations—the archive of feelings—of the creators of such messages. Are the messages designed to give hope to the viewers *and* to the creators themselves? Are the stories offering viewers ways to endure their present suffering *and* helping creators to overcome their own past traumas? To expand and move such reflections from the individual (microscopic) to the structural (macroscopic) levels, can the social world get better for LGBTQ individuals and communities under current symbolic, ideological, structural, and cultural systems of heteronormativity? What does a queer world look like?

The Question of Queer World-Making

To create a queer world, LGBTQ individuals and communities must transcend by creating spaces for sexual intimacy and offer new possibilities of identity, social relations, intelligibility, sexuality, and culture in which the heterosexual couple is no longer the normative, privileged, and taken-for-granted prototype (Warner, 2002). Yep (2003) extended this notion beyond the frames of heterosexual privilege:

> Queer world-making is the opening and creation of spaces without a map, the invention and proliferation of ideas without an unchanging and predetermined goal, and the expansion of individual freedom and collective possibilities without the constraints of suffocating identities and restrictive membership. (p. 35)

In this sense, a queer world is not necessarily coherent and neatly delineated because it embraces all the intersections and complexities of human relations. In many ways, it is plural—queer worlds—and includes liminal spaces for transformative symbolic, collective, and material possibilities.

Although the videos appear to give voice to, offer hope, or even queer the experiences of gender and sexual minorities, they also normalize and maintain the assumption that LGBTQ people are ultimately similar to their heterosexual counterparts. For example, as previously discussed, personal and professional success—engaging in certain types of normative intimate relationships, establishing economic productivity, and procuring material achievement—are presented in terms characteristic of homonormativity (Yep & Elia, in press) and Duggan's (2003, p. 50) "new homonormativity."

The videos offer glimpses of the promise that queer world-making can occur, but frequently they collapse into propensities for assimilation(s), normativities, and homonormativity. Our reading of the stories suggests two themes associated with the question of queer world-making: challenging the traditional heteropatriarchal family, and contesting and queering surveillance of non-normativities.

Challenging the Heteropatriarchal Family?

The presence of LGBTQ individuals can challenge the traditional heteropatriarchal notion of family, and its cornerstones of male supremacy and heterosexuality. This is evident in our videos, as when one of the four African American gay men in AJ541's video discusses growing up with the women in his family:

> I wanted to come out of the closet and during that time, it was very, you know, um, traumatic for me because my family is very anti-gay, you know, very homophobic. I come from a family of women that believe that . . . homosexuals . . . molest children . . . [T]hey've always been anti-gay . . . the ironic thing about it is that my father was bisexual so clearly my mother is kind of fucked up.

His revelation has elements of queer world-making, yet it reiterates racialized sexual stereotypes. A female-centered household disrupts patriarchal notions of family and the bisexuality of the father holds promise to disrupt heterosexual conceptions of family. However, it also reinforces stereotypical conceptions of African Americans as homophobic and African American families as unstable.

Similarly, other videos complicate conceptions of family and patriarchal power. In his story about growing up in a Syrian Muslim Pakistani household, lecercle01 says:

> [M]y dad had made it clear before to my family that if it turned out that any of his kids were gay, that he would, ah, walk out on the family, 'cause he can't bear . . . the shame, I guess, or what the rest of the community would think about us as [a] family and him as a father.

However, after finding out that his son is gay and "outing" him to the rest of the family, the father came around "within really a day or two" after coming to the realization that his wife and daughters did not object to his son's homosexuality. This narrative displays elements of queer world-making—undermining patriarchal power and the importance of the voices of the women in the family—and reification of racialized gender and cultural stereotypes—the image of a homophobic and dominant Muslim father in a shame-avoiding and honor-driven culture.

Queering Surveillance?

Surveillance, in the context of the stories, refers to the process of being followed, watched, or observed in order to police certain non-normative behaviors associated with gender performance and sexual identity. The paradox of this theme lies in the complexity of being outed or challenged about one's identity, and the ways this process can create a space for queer world-making.

In the videos, families and other social institutions scrutinize and regulate behaviors that challenge heterosexual ideals and gender binaries that uphold normative understandings of gender and sexuality. However, such scrutiny and regulation is complex and can produce unintended—and queer—consequences. For example, lecercle01 said:

> So rather recently, ah, my dad sat me down [and] said [that] I was, ah, seen in a gay bar, which I've never been to . . . in my life, but I am assuming he means, ah, this "True Colors" conference [a meeting for sexual minority youth and their allies] that I went to at the University of Connecticut, . . . it just kind of devolved into this really contentious argument that went under the assumption that I was gay.

The narrator's participation in an event focused on queer world-making—an academic conference for LGBTQ youth and their allies—implicates him in his queerness, which creates an atmosphere of shame and blame. However, an interesting turn of events takes place:

> He had told my mother, he had told my sisters . . . he had basically told me that I was a failure of a son, he had failed to raise me properly . . . but within really a day or two, he came around, it was like nothing happened. My mom proved not to care, my sisters especially couldn't give less of a shit, ah, and he came around to it.

The way his father outed him produced some unintended consequences: He is now open about his sexuality, his mother and sisters do not reject him, and his father accepted him quickly, thus creating a more positive and open communication environment in his family.

However, others have different experiences about being outed. For example, victoriaray shares her story of physical violence and abuse, discussed earlier, which speaks to the real dangers of coming out. Victoriaray discusses the journey of coming to terms with her sexuality, the ongoing healing of the relationship with her mother (which at the time was uncertain), and the process of forgiveness. Her journey disrupts normative ideals that one must be in good standing with one's parents and problematizes narratives of progress and the promise of happiness, creating a space for queer potentiality. Queer world-making is about creating new ways of seeing, acting, and being, and using these new inventions to move through, and perhaps change, the social world.

Audience-Generated Media and Queer World-Making

Queer world-making is not an easy and straightforward endeavor. The proliferation of UGC in our current media environment has the potential to open up and create new spaces for multiple perspectives, visions, and voices to be heard, become intelligible, and to form new affinity networks. UGC is, in a number of ways, about producing new theories and new forms of engagement with the social world. This potential for greater democratic engagement, however, exists within the larger cultural, ideological, and economic systems of neoliberal capitalism, which promote the notion of individualism at the expense of community. This is evident in the IGB video postings. The almost exclusive focus on the individual and the microscopic level of interaction in the stories (e.g., self-acceptance, interpersonal relationships) often overlooks and neglects the larger structural and macroscopic factors (e.g., differential legal treatment of LGBTQ people in society, erasure of gender non-conforming and sexual minority people in history) that sustain and perpetuate homophobic and heteronormative violence in our culture. UGC media can—and in our view, should—be a critical component of queer world-making. To do so, however, we must imagine life beyond normativities and the individual.

Note

1. Gust thanks Dr. Rebecca Lind for her flexibility and understanding; Dr. Donald Arquilla for his insightful clarity; Bill Heter for his gentle and ongoing support and Rocky for his vocal and enthusiastic presence; and Yogi, my affectionate and inquisitive Pomeranian companion, who knew when to remind me to take a break during the hectic period of preparation of this manuscript.

References

AbleTraveler. (2010). *It gets better - gay disabled man* - Portland Oregon [Video file]. Retrieved December 8, 2011, from http://www.youtube.com/watch?feature=player_detailpage&v=DRn4Ueaf2pw

Ahmed, S. (2010). *The promise of happiness*. Durham, NC: Duke University Press.

AJ541. (2010). *It gets better project - a message to gay youth* [Video file]. Retrieved December 3, 2011, from http://www.youtube.com/watch?v=Ir1EempCAmo

Benjamin, W. (1968). Theses on the philosophy of history. In H. Arendt (Ed.), *Illuminations* (H. Zohn, Trans.) (pp. 255–266). New York, NY: Harcourt, Brace, & World.

Berlant, L. (2011). *Cruel optimism*. Durham, NC: Duke University Press.

blackgayandjewish. (2010). *It gets better project-black gay Jew* [Video file]. Retrieved December 1, 2011, from http://www.youtube.com/watch?v=tLJ6tgg-rb0

Browne, K., & Nash, C. J. (2010). Queer methods and methodologies: An introduction. In K. Browne & C. J. Nash (Eds.), *Queer methods and methodologies: Intersecting queer theories and social science research* (pp. 1–23). Farnham, England: Ashgate.

Caruth, C. (1996). *Unclaimed experience: Trauma, narrative, and history*. Baltimore, MD: Johns Hopkins University Press.

Cvetkovich, A. (2003). *An archive of feelings: Trauma, sexuality, and lesbian public cultures.* Durham, NC: Duke University Press.

dudessforrent. (2010). *It gets better* [Video file]. Retrieved December 1, 2011, from http://www.youtube.com/watch?v=D4vWk5aXXYs

Duggan, L. (2003). *The twilight of equality? Neoliberalism, cultural politics, and the attack on democracy.* Boston, MA: Beacon Press.

Foucault, M. (1978). *The history of sexuality, Vol. 1: An introduction* (R. Hurley, Trans.). New York, NY: Vintage.

Halberstam, J. (1998). *Female masculinity.* Durham, NC: Duke University Press.

Halberstam, J. (2010, November 20). It gets worse... *Social Text* [Blog post]. Retrieved November 29, 2011, from http://www.socialtextjournal.org/periscope/2010/11/it-gets-worse.php

Irving, D. (2008). Normalized transgressions: Legitimizing the transsexual body as productive. *Radical History Review, 100*(Winter), 38–59.

It Gets Better (n.d.). *What is the It Gets Better Project?* Retrieved June 29, 2011, from http://www.itgetsbetter.org/pages/about-it-gets-better-project/

lecercle01. (2010). *It Gets Better* [Video file]. Retrieved December 1, 2011, from http://www.youtube.com/watch?v=Wj8zBtcgTjA

Love, H. (2007). *Feeling backward: Loss and the politics of queer history.* Cambridge, MA: Harvard University Press.

Ngai, S. (2007). *Ugly feelings.* Cambridge, MA: Harvard University Press.

Puar, J. (2010, November 16). In the wake of It Gets Better. *The Guardian.* Retrieved November 29, 2011, from http://www.guardian.co.uk/commentisfree/cifamerica/2010/nov/16/wake-it-gets-better-campaign

Puar, J. (2012). Coda: The cost of getting better: Suicide, sensation, switchpoints. *GLQ: A Journal of Lesbian and Gay Studies, 18*(1), 149–158.

Savage, D., & Miller, T. (Eds.). (2011). *It gets better: Coming out, overcoming bullying, and creating a life worth living.* New York, NY: Dutton Press.

TheKairielise. (2010). *It gets better* [Video file]. Retrieved December 3, 2011, from http://www.youtube.com/watch?v=4_qoIkuAHWI&feature=related

victoriaray. (2011). *It gets better* [Video file]. Retrieved December 3, 2011, from http://www.itgetsbetter.org/video/entry/it1f6m61u5o/

Warner, M. (2002). *Publics and counterpublics.* New York, NY: Zone.

Yep, G. A. (2003). The violence of heteronormativity in communication studies: Notes on injury, healing, and queer-world making. In G. A. Yep, K. E. Lovaas, & J. P. Elia (Eds.), *Queer theory and communication: From disciplining queers to queering the disciplines* (pp. 11–59). Binghamton, NY: Harrington Park Press.

Yep, G. A. (2009). Queer theory. In S. W. Littlejohn & K. A. Foss (Eds.), *Encyclopedia of communication theory* (Vol. 2, pp. 817–821). Thousand Oaks, CA: Sage.

Yep, G. A., & Elia, J. P. (in press), Racialized masculinities and the new homonormativity in LOGO's *Noah's Arc*. *Journal of Homosexuality*.

CHAPTER NINE

Black Penis/White Phallus: The Virtual Outsourcing of Perverse Labor (Or, the Cuckold Fantasy as Colonial Encounter)

Diego Costa

In *Black Skin, White Masks*, Frantz Fanon (1967) stated that the white man wants the world all for himself. This is a wish, and a project, for masterful predestination that he scripts through enslavement, an enslavement that highjacks the Symbolic as an exclusively white code (of conduct, of humanness) against which the black man will always fall short. We could say that Fanon's post-colonial theory mourns this constructed schism between the always-white Symbolic and the lived experience of the black man, which can only be legible as failed yearning for an access that is inexorably denied.

We could perhaps call this alienation a *queer* feeling (in both the original and post-academic sense of the word). Fanon argued that white eyes are the only real eyes (p. 116), a predicament that walls the black man in, fixing him into place (p. 117), locking him "into the infernal cycle" (p. 116), and forever leaving "a world—a white world—between you and us..." (p. 122). The outsideness of the black man is part and parcel of the white world, whose very logic depends on the mapping of otherness as inextricable from certain bodies. This world, which is constructed as "the only honorable one," and which bars the black man from "all participation" (Fanon, 1967, p. 114), requires maintenance—and a little perversion. Its constant reiteration, justification, and otherization all involve actual *labor*. In this chapter I explore what the cuckold fantasy as seen in pornographic produsage reveals about the gap between the already white Symbolic and black masculinity, and how the black man is interpellated when claims of the white man's ideality cannot sustain themselves as true performatically. Fanon cited a friend who was a teacher in the United States, "The presence of the Negroes beside the whites is in a way an insurance policy on humanness. When the whites feel they have become too mechanized, they run to the men of color and ask them for a little human sustenance" (p, 129) This relationship of outsourcing, when blacks become representatives, spokespeople and laborers of perversions, or non-normative activity more generally, that are either white or of the subject *tout court*, is most evident in slavery, but remains a staple of American culture; whether blacks are mining for metals, weaving cotton, forging steel, or providing misogynist club anthems for all to dance to, the principles remain quite similar (p. 130). Fanon explained:

> The civilized white man retains an irrational longing for unusual eras of sexual license, of orgiastic scenes, of unpunished rapes, of unrepressed incest. ...Projecting his own desires onto the Negro, the white man behaves "as if" the Negro really had them. When it is a question of the Jew, the problem is clear: He is suspect because he wants to own the wealth or take over the positions of power. But the Negro is fixated at the genital; or at any rate he has been fixated there. (1967, p. 165)

What can the scene of cuckoldry, interracial (as I will argue) even when it's not, say about male blackness' undergirding, enacting, performing and guaranteeing the machinery of constitutive *difference* (racial and otherwise)? How does black masculinity, or the white fantasy thereof, work as a prosthetic guarantor of the Symbolic through the pornographic?

Amateur porn is readily available online on innumerable Web sites that categorize content by an extensive list of genres or user-generated tags, such as "group sex," "gangbang," and "interracial." Many offer users the opportunity not only to upload their own videos, but also to comment on the material posted by others. Material tagged as "cuckold" normally features the following actors: a white heterosexual couple interested in spicing up their relationship or seduced by the financial reward provided by the pornographer behind the camera, and the outsider male, often referred to as the "bull," invited to have sex with the white wife in front of her husband.

The Web site CuckoldFantasies.com (2012) defines the cuckolded husband as the "beta male" and the wife as "hotwife," and offers a list of different types of bulls, or studs: The Younger Man, The Pool Boy, The Ex-Factor, and The Black Male, described as "usually very hung; much more than the cuckold husband." It then tells us that "black men love fucking white women and will do this with great pleasure when given the opportunity." But how is it that cuckoldry begets interraciality so easily (most of the videos tagged "cuckold" feature black bulls and white couples)? How is it that blackness is the color of the bull? Why do we read interraciality as a sign of cuckoldry (some videos tagged as "cuckold" simply feature a black man and a white woman, with no husband physically in the scene)? And what might the intrinsically interracial cuckold porn reveal about a much more pervasive, and less explicitly pornographic, dynamic between omniscient whiteness and laborious blackness—what I am calling *the virtual outsourcing of perverse labor*?

For Fanon, the black man wants to be white, and his desire to go to bed with a white woman proves it, either through its erotic investment or as a lust for revenge mapped onto her body, or both (1967, p. 14). We can also see in Fanon that the black man's yearning for white attention is an attempt of liberation, in that he finds himself "an object in the midst of other objects" (p. 109), a situation that brings his blackness to the surface. Of course one doesn't have to be a black man to demand outside attention as a way of feeling loved (or at least existent), or to be an object among other objects, but here Fanon's adverb choice is quite telling: "I turn beseechingly to others" (p. 109), suggesting a sense of fulfillment in living up to his expected subservience. If, for Fanon, the black man *works,* beseechingly—the white man being the symbol of capital whereas "the Negro is that of labor" (p. 133)—and the black man has no access to the (white) Symbolic, the scene of (interracial) cuckoldry would be perhaps too seductive an interpellation as it glazes the complete instrumentalization of the black man by the white man with the ultimate, if fleeting, reward: the white woman's body.

Let us not forget that it was "in the train" ("running train" being another term for engaging in a group sex scenario in which one passive partner gets penetrated by various men), through the constant gaze of "others," that Fanon was most aware of his blackness (p. 112). The moving train is the site where Fanon has taken himself "far off from [his] own presence" (p. 112) and becomes object. Interestingly, he associated this awareness of physical boundedness (this boundedness is literalized as he decides to assert himself as "a BLACK MAN," p. 115) with being reduced to color (the "corporeal schema" has been replaced by "a racial epidermal scheme," p. 112), and linked it to a sense of castration. In the moving train the black man becomes thing while feeling maimed. In the sex train he becomes a penile prosthesis for the white man's phallus:

> What else could it be for me but an amputation, an excision, a hemorrhage that splattered my whole body with black blood? ...All I wanted was to be a man among other men. I wanted to come lithe and young into a world that was ours and *to help to build it together.*" (pp. 112–113, emphasis added)

The Virtual Outsourcing of Perverse Labor

Lucien Israël's (1996) notion of perversion is that of sexual manifestation that escapes mythic normativity, that is, virtually all de facto sexuality. Perversion is a space of refuge for the subject, in which the pervert produces a closed field to put his fantasy at play in a way that he, or she, can control its every element, dynamic and narrative. This mechanism of defense bodes well and works in tandem with the reification of the naturalized myths about self and other (these myths sustain the subject's sense of existential cohesion, and are reflected in racist, colonial, sexist, and homophobic discourse). It is worth noting that the pervert creates this sealed-off world for himself away from a gaze that demands normative (sexual) practice, a world he can visit and convince himself—because the inhabitants of that exterior world have no access to the insular one—that one is no pervert. As the digital becomes such a pervasive part of everyday life (where normativity stakes itself), it becomes increasingly hard to keep the porousness between such worlds from coming to the surface. Within the scene of cuckoldry, too, these worlds can collide, or at least come together temporally, because the action tends to take place within the confines of the ordinary middle class spaces that are purposefully invaded by the other-as-pervert, the bull, who is hired to do the dirty job from which the white man excuses himself.

It is no accident, then, that the hotel room is a recurring setting for cuck-oldry as enacted and uploaded by users. The hotel room is, after all, a transient space normally located in a city where one doesn't belong. It appears as a perfect non-space for the pervert to perform his scene without allowing for one world to contaminate the other completely. "My cuckold wife fucked this black fucker at the nastiest shithole motel while I filmed," is how Xtube user LT42011 (n.d.) highlighted the setting of the scene in his video *Cuckold Fucked in Motel*, for example. The quality of this user-generated video leaves no doubt as to its amateur status, which is what seems most alluring about the video (besides the appearance of the cuckold) for commenter Matt4uat20009, who praised the video's realism and then went on to say that its best part is "the few seconds we see the cuck in the mirror filming the bull pounding the wife." He continued:

> After many years of monogamous marriage my wife and I started having sex with other men. Soon the roles emerged with me usually the cuck if the guy was an alpha male. As in this vid the cuck is clearly in charge of the whole scene. The bull is an accessory.

Melcuck's (n.d.) video *Fucking Her BBC Raw for the First Time* is also set in a hotel room. In the description Melcuck apologized for not showing the wife and the bull's faces "for obvious reasons," and pointed out that this is the first time a bull penetrated her without a condom. One of the commenters, hukindian1980, taunted him: "i think her pussy belongs to him now?"

Israël (1996) claimed that the pervert, contrary to popular belief, is no transgressor of laws—he actually respects the law, and makes good use of it. He knows something that the normative Subject doesn't (that anxieties about the white woman's rape can be managed by the white man's pre-emptively scripting it and casting the black man to perform it, for instance). With the increasing popularity of social networking sites and mobile technology, the subject can effectively place the pervert world and its putatively non-pervert counterpart in simultaneous planes (pornography consumption at work, or on the go)—without ever completely leaving it. He is able to live in both worlds simultaneously with perversion safely hovering over the presumed normalcy of everyday life like a fifth season, or an eighth day of the week. The ease and ubiquity of digital technology affords the cuckold even more layers of control, because he can edit footage as he wishes, contextualize the images in the descriptions of the videos, comment on videos uploaded by others, and receive what might be considered "safe" acknowledgment from commenters—safe because it doesn't expose his identity as a pervert.

At the core of perversion, however, is exclusivity. Perversion lies in the necessity to over-determine the object of desire and the kind of relationship the Subject has to that object. Hence Israël, ironically, declared monogamous marriage the ultimate perversion. In a sense, this levels the playing field, perversion being the domain of desire more generally, and thus, the domain of the human *tout court*. The man who adores the Negro is as sick as the man who abominates him, as Fanon (1967) reminded us. As a prominent clause of a well-scripted arrangement (Israël spoke of the pervert's contract) the bull can only have the woman for a short amount of time. The representative of the phallus, after all, can only last so long in its delirium before it is found to be lacking. The temporal limits of the bull's prosthetic work—his limited usability, or worth—is crucial to the arrangement. Much like the relation between colonizer and native, in which the latter might have specialized talents or useful knowledge but the former holds the indomitable spirit and intelligence to put all the parts to work, the bull may be seen as having the more powerful physical equipment, yet he only holds the stuff out of which greatness is made, not the ability to put the scene together. This is perhaps katsbitchboy's sentiment, who commented on biguy60123's *Cuckold, Creampie Cleanup* (n.d.) video: "White hubbies should, IF they LOVE her, eagerly give their wives to superior Black men!" The superiority seems completely contingent on physicality and hugeness. Commenting on the same video, viper_99999 said he is "amazed at the size of the huge cock."

Produsing and Colonizing the Bull

The cuckold scene is a colonial encounter in that it safeguards the imagined materiality of the white man, who outsources the material representation and performance of his phallus, the phallic labor of the penis, to the bull. Although this move may be seen as a kind of aid to the bull who is granted access to the white woman, albeit for a short time and probably just this once, the bull's pleasure is mostly incidental to his interpellation as both laborer and white man's remote—but not too remote—prosthesis. The black bull cannot speak; he is interpellated and he complies. The bull's sexual labor is always in the name of, and bound by, the white man's fantasy. In a rare instance, kinggenton must mark himself as black when he says that "this is my fantasy! (im black)" in response to HKHORNY's (n.d.). *Sexy Asian Woman Cuckolding with Black Guy* video. One is left to wonder whether kinggenton's fantasy is to have his (Asian) wife cuckold him as in the video or to be put in the place of the black bull himself. The bull is visibly bounded as a priapic penis for the always already white phallus, which, in Lacan's (2007) canonical words, is a signifier without signified. Here the signifier finds its ephemeral haze of a signified in its phallic (black) prosthesis. "The Negro," said Fanon, "is a toy in the white man's hands; so, in order to shatter the hellish cycle, he explodes" (1967, p. 135).

There is something excessive about the bull, whereas the husband tends to represent the regular Joe who may be consuming, producing, or *produsing* (Bruns, 2008) the video or commentary. The bull in these videos is likely to be stronger, bigger and younger than the husband, conforming to what we could call a hyperbolic rendering of masculine ideality for the Western collective unconscious. The bull's massiveness and anti-intellectual function (he simply fucks) props up white man's power and superiority, releasing the cuckold from his duties to perform a phallicity he could only ever fail to translate physically.

In the cuckold scene the husband sometimes joins the bull in having sex with the wife, only to underscore the physical superiority of the bull; the wife is always more interested in the bull. At other times the husband is forced to watch the scene like a powerless, perhaps humiliated, spectator, sometimes wearing a chastity belt type of device that prevents him from even masturbating to the spectacle. The cuckold can also be asked to arrive at the post-coital scene and clean up the damage (to the woman's body) done by the bull. There are also instances when the husband is forced to submit to the bull himself, such as by performing oral sex on him. Even here, in his position of feigned subservience, we could say, as did *Xtube* commenter john2006 in response to wimp2black's (n.d.) video *Cuckold 1*, that the white cuckold is simply "having your dick and eating it too!"

As delicate as everyone's position in the room is in the scene of cuckoldry, however, their locations are also inextricably guaranteed to spring them back to the normalcy of the mythically monogamous heterosexual couple once the bull empties himself and clocks out. There is apparently "no coercion, only mutual assistance" (Fanon, 1967, p. 131). Fanon claimed that the white man's anxieties about the possible escape of the black man (perhaps taking "something" with him) eventually lead to the black man's "free acceptance of discipline" (p. 131). The cuckold scene, which brings together the fragility of race, class, gender, and sex constructions, simultaneously works to prove the staunchness of their significations. As Xhamster.com user Everdark put it when commenting on SexyCassie's appropriately titled *Wife Cheating with Stranger in Motel* (2011), "Fucking with black ain't cheating." For renegadebob "It is only cheating if her husband does not know."

If distances have never been so excerpted materially, their symbolic borders have also never been so guaranteed in their very proximity, for this bouleversement of the mathematics of the heterosexual theater [*1 (white) man + 1 (white) woman*] has only been made possible by the acceptance of the very solidity of the positions occupied by its actors—and hypercontrolled by at least one of them (never the bull or the woman). Even in instances when the woman seems to have uploaded and described the videos, there is the sense of the cuckold's permission preceding the act(s). The wife's feisty humiliation of her husband, in the putatively defiant act of posting evidence of cuckoldry as a performance online is ultimately just that—an act. Tara_Tainton (n.d.), for instance, posted an elaborate description for her video *Knocking My Small-Dicked Cuckold Boyfriend Down—Part 3*, which she directed at the cuckold himself:

> I'll put on a little show for you...let you peek beneath my bath towel and remember what you used to enjoy...BEFORE you became my cuckolded lover shrunken down to the size that suits you, your small penis, and your other inadequacies best .... I intend for you to become my sexual slave, so I need you to...become a great, big satisfying dildo!

Notice how size becomes the unit of reference for both idealization (the bull is a "great, big satisfying dildo") and debunking (the cuckold is "shrunken"). The entire affair, however, is, as she worded it, "put on" as a show "just for" the cuckold, who also remains the only gaze worth fighting for, the only gaze there is.

It is worth noting that more than many other porn niches, cuckold porn seems to necessitate redundant verbal explanation, or reassurance, of the narrative premises. Professional videos often mimic amateur efforts, and to reinforce this illusion they often include interviews with the white couple completely clothed, sometimes in a public place, setting up the scene before the bull shows up to do his job. The couple usually tries to convince us of their absolute heteronormativity up to that point, explaining in detail how this is the first time they have ever done this, telling us how long they have been married, and so on (building up the massiveness of their normativity to add contrast and drama to the bull's foreboding piercing of it). In an interesting reflection of the professionals' practices of mimicking user-generated videos, the amateur videos themselves also often include such an explanation—in this case, alongside the video's details and technical specs. Melcuck (n.d.), in his video *Cuckold Double Cream Pie*, which has received hundreds of thousands of views on Xtube, framed the scene this way:

> My wife (43, married 21 years) fucked this bull on our bed and made me video everything. I had to make her cum with my tongue before he arrived so he could slide straight into her. You will see me guide his big meat into her (without a condom) then him pounding her and then breeding her with a huge load (watch her squeeze his balls as he shoots). Then they make me fuck her full of his cum while the Bull videos until I shoot inside her on top of his load. You can see it running out of the dirty bitch. Turn the sound up and listen to me being truly cuckolded by these two. If you are in Melbourne and have a big cock get in touch, you may get to meet her.

It is interesting that Melcuck, who presumably directed the scene, gave specific directions to the anonymous viewer as to how he (and it is presumed to be a he) should consume it, from encouraging his gaze to focus on certain elements of the image to the teasing suggestion that he may even get to play the bull in a future real-life re-production of the scene, provided he "[has] a big cock." The viewers seem to follow directions well, their comments reflecting their attention to what Melcuck had highlighted. "Love how you guided his fat headed dick straight into her sloppy cunt," said joejackk. And addictedmasterb8er said, "the best part is how the cuck is holding the bulls cock and guides it into her wet pussy…"

This interpellation of the word to support the visual fantasy can seep into the image itself, or the white woman's body more specifically. The "wife-writing" tag involves words written on the woman's skin, usually around her genitals or her mouth (sometimes she will hold a handmade sign). The words are scribbled with lipstick or markers and usually underline the wife as an object in an asymmetric transaction between men (she is borrowed, not exchanged): "my black owned wife," "black dick only," and "black cock slut" are examples (Cuckold Space, n.d.).

Prosthetic Penetration:
Working Without Breaking, Laboring While Resting

The bull is a beast, pre-civilized, primitive, a machine (less a smart computer than a robust gadget). The cuckold's black man is less than human because of his *functionality*: he is a mechanical bull, a machine requiring outside aid to be able to work. He is hardware. If the bull is not fighting against the machine but is *a* machine, the cuckold is *the* machine that calls him into being. The cuckold delegates his sexual labor, which in its idealized phallicity will always fail, in order to excuse himself from going beyond representation and actually attempting to prove, in the flesh, the existence of the phallus. This symbiotic and retractable distance reflected in such a complete instrumentalization of the black man exposes the inadequacy of the white man's penis, or the penis *tout court*, as a supposed stand-in for the phallus. Disguised behind the self-scripted helplessness of the cuckold is his passionate—fervor is, after all, the weapon of the impotent, according to Fanon (1967)—and pre-emptive staging of a tragedy: a tragedy (of betrayal and absence, of phallic transvestism, really) that would otherwise come as an unwelcome surprise.

This arrangement reveals heterosexual sex as a *practice* as (un)necessary nuisance, or mere labor, of a *symbolic* system whose circuitry of *desire* always manages to channel itself elsewhere. The scene of cuckoldry points to the breaking down of heterosexual logic (based on the penile mapping of the phallus) while inhabiting a tremendous space of liminality in which two men have never been so sexually close yet so heterosexually certified. After all, woman is being beaten. Her beating occurs in a context of highly divided labor strategically abdicated by one person and carried out by the other. The white man's penis is, thus, as Beatriz Preciado (2000) would put it, a mere dildo. With luck (and some perverse outsourcing), as Tara_Tainton (n.d.) would have it, a "great, big satisfying dildo."

Following Foucault's notion that the best resistance is not a fight against prohibition, but a counter-productivity, Preciado developed a manifesto toward a *contra-sexuality*, which she described as a theory of the body outside oppositions (male/female, masculine/feminine, hetero/homo). Contra-sexuality aims not for a symbolic political intervention, but a re-positioning of the subject vis-à-vis sexuality, which is defined as a technology, the heteronormative social technology. The elements of the sex-gender system (man, woman, homosexual, heterosexual, transsexual) are machines, products, tools, gadgets, prostheses, networks, applications, programs, connections, fluxes of energy and data, borders, constraints, designs, logics, accidents, detritus, mechanisms, and diversions. This context, in which human nature is an effect of social technology that reproduces in the body the equation "nature = heterosexuality," follows the penis as the only *mechanical* center of the production of sexual drive.

Preciado reminded us that the penis is a genital organ whereas the phallus is a privileged signifier that represents power and desire, and sanctions access to the Symbolic order (the fact that not all penises are made equal is not part of her project). Hence the dildo occupies a strategic position between One and Other, white cuckold and black bull. The dildo denounces the pretension of the penis in passing for the phallus. It is a sexual technology that occupies a tactical place among technologies of repression and of pleasure. Preciado also noted that in the market of heterosexual sex toys, men are offered the totality of the woman's body (in the shape of a sex doll), whereas women are left with just one part of man's body (in the shape of the "realistic dildo"). This makes explicit the asymmetry between men and women in their access to the sexual. It is also relevant to a critical reading of the cuckold scene in the way the entire body of the black bull—and body is all he is—is transformed, in a counter-synecdoche of sorts, into one big dildo controlled by the other man in the room: the (white) cuckold who sanctions access. In a sense, the cuckold simultaneously utilizes the bull for his "great, big satisfying" potential and feminizes him into the scene, where anyone else is a (sex) doll. Similarly, the white man in Fanon snares him through an interjected Symbolic he can only play *for*.

In this sexual outsourcing move which turns the black man into a functional totality, the white man relinquishes from performing his phallus through its supposed sign (the penis) in a very controlled way that assures its unquestionable competence and productivity. As in Eldridge Cleaver's (1999) schema of heterosexuality divided into four categories of people (white man and woman, black man and woman) and their functions, the outsourcing—which may feel phallic to the bull but is only penile to the white man—is also always already there. It only becomes fleshly in the enactment of the scene. Ironically, in divorcing the phallus from the penis that is attached to himself organically in the *dildofication* of the black bull's body, the cuckold manages to protect his penis and the phallus (which is always already his) from the risks that its performance occasions. The move may leave the cuckold's phallus unscathed but it does reveal Preciado's (2000) claim, drawn from Judith Halberstam, that dildos expose the fact that penises are nothing but dildos themselves. As Tara_Tainton (n.d.) put it, however, it takes outsourcing to the black bull for that dildo to gain "great, big satisfying" status and, thus, translate its supposedly intimate relationship to the phallus into sexual performance.

Because the black bull-as-dildo intervenes in the naturalized logic of the penis as phallus and the phallus as white, the cuckold's elaborate script works to divide up the labor as a way to save the heterosexual logic of nature from displaying its artifice—even if it ultimately fails. As long as the phallus remains white, the penis can be black. In fact, the penis may *need* to be black for the phallus to remain white.

For Preciado, almost all new technological developments of the 20th century were some sort of prosthetic supplement of a so-called natural function. Instead of being an artificial substitute for a living organ, the prosthesis became a supplemental technology of enhancement: the telephone (communication), television (hearing and vision), cinema (dreams), touch-screen (haptics), the personal digital gadget (masturbation). It's thus hard to separate the prosthetic from the bodily. In an organic-inorganic web of individual and collective continuities (Lacan's [2007] metaphor of the unconscious as a hoop net comes to mind), everything becomes prosthesis: homes, cars, communication systems, radio, telephone, TV, computer (Colomina, 2007), and, why not, the construction of racial/national/sexual others.

Foster (2006) called attention to modernism's "double logic of the prosthesis" (p. 109), which involves both extension and constriction. Foster related a cultural imaginary traumatized by the mutilated bodies of World War I to the prosthesis, suggesting it as a way "to make over this body-ego image damaged in reality and representation alike" (p. 110).

When the videographer is a disembodied character eliciting the cuckold's humiliation, we can further see a mise en abyme network of prostheses starting with the voyeurism of the consumer or produser all the way down to the performing mechanical bull. Foster recognized in Freud's concept of binding (*Bindung*) and unbinding (*Entbindung*) the ghostly mechanism of defense: "If the ego is breached in situations of shock, the process of integration is threatened: hence the imperative to shield the ego before its breaching, or to shore it up afterwards" (2006, p. 115). We can see the bull as the shield that enables the white man's phallus to work without breaking, to perform without shame, to labor while resting. Whether he is behind the camera or gazing at the scene from the couch, his own risks are removed from the action. In zen_sun84's video ****** *Cuckold* (n.d.) the titular character is literally bound to his chair, with a ball in his mouth disabling his speech (why speak when the overdeterminedness of the scene has been so perfectly coded?). He is at once supposedly humiliated by the inability to speak or move and receiving a blow job from the wife, whom the bull penetrates. Here we may ask whether the wife herself isn't the remote prosthesis that works to shield the cuckold's desire, and mediated feeling, of being penetrated by the bull himself.

This dynamic of eruption and resistance dramatized in the cuckold scene as the aggression represented and performed by the bull serves to further reify, as violently as possible, the woman as object and the bull as instrument, and re-renders the omniscient subjectivity of the cuckold subject. At the same time, and it is fundamental to recognize this binding and unbinding not as polarities but simultaneities, the cuckold might manage his own penetration through the proxy of the woman without ever having to endure the labor, and symbolic devaluing, that the literal non-mediated execution of such an act would produce. We can call it remote penetration, prosthetic penetration, or vicarious penetration. Whatever the label, it works without inaugurating a difference between passive and active, for the bull is additive to the cuckold in a division of labor that preserves the *activity* of both (yet only reduces one of them to it): the literal action of the bull and the abstract, intelligent and designing sagacity of the cuckold. The bull is the "inchoate terrain needing the skill, sense and vision of the colonizer to be brought to order," or put to good use (Dyer, 1997, p. 165). The scene of cuckoldry, along with its pre- and post-production and produsage (in the form of scripting, casting, uploading, disseminating, and managing comments) thus brings together the binding and unbinding of the white male subject, who remains unscathed: the self is made safe(r) through the contained violence of the Other toward another Other, as the *jouissance* provoked in the self is of no detrimental consequences to himself.

Borrowing, Sampling, Losing

The interracial cuckold fantasy in porn inhabits the very limit of white heterosexual patriarchy because it happens in the instant before its traumatic undoing; it is always almost about to happen, and yet it never will. There is no real jealousy on the part of the white man because there is a certainty of property, a certainty that enables the very enactment of the scene, and solidifies it. The scene is set up in a way to assuage anxieties about loss (of property). The scene never happens in a space belonging to the bull (none of the videos ever show the bull as having a home, or a wife), but rather in a space controlled by the middle class white couple. There is never the slightest hint of an actual *undoing* of the roles, because that could certainly trigger the rapid and effective undoing of the manifested setup itself and, thus, the immediate withholding of the white woman from the black bull.

The idea of loss of (female) property is so safe as solely an idea that it is literally just *entertained*, like a fort/da children's game, which allows the child to hate the object being thrown and revel in the certainty of its timely return. Whereas Rubin (2004) deemed the white woman the most valuable prize in the exchanges between (white) men, Derrida linked the very principle of jealousy to "the primitive passion for property" (1993, p. 3) in his discussion of life itself as a kind of property one can share, give, or waste. In cuckoldry users produce, produse, and reiterate life as belonging to the white man who shares the white woman in a kind of masturbatory tease, a hyper-supervised borrowing made possible by the knowledge that she is structurally, symbolically, and forever tied to him—and if ever beyond his touch, never out of his sight. In contrast, the black man is absorbed in his own prosthetic doll-like totality. The white woman may work as a gift to establish kinship bonds between men, but she remains *someone else's* gift that the black man is allowed to *sample*, not to *have*. The black bull functions as what Derrida (1993) might have called an interpellated arrivant, invited by the white man himself so that, unlike the absolute arrivant who surprises the host, his condition isn't unsettled. Derrida described the absolute arrivant as the non-guest who calls "into question, to the point of annihilating or rendering indeterminate, all the distinctive signs of a prior identity, beginning with the very border that delineated a legitimate home and assured lineage, names and language, nations, families and genealogies" (p. 34). We can imagine the possibility for the black man to become an absolute arrivant and damage something other than the woman's body, but we can also see how the scene of cuckoldry is rigged to prevent such an occurrence.

Because of this, cuckolding scenarios featuring a black anonymous bull enable produsers, participants, and spectators to revel in what colonial theory would call "primitive" pleasures inhabiting, or spawning from, the registers of racism, classism, colonialism and other perversions (as we have seen, relationships of extreme control and dehumanization are inherently perverse).

Dyer (1997) found a similar scene of re-solidification of the Other as Other in his study of boxing and adventure films in colonial settings, with stars possessing a built body as telling exceptions of a culture that, until the 1980s, reserved the display of non-clothed male physicality to the non-white body. Following Eldridge Cleaver's (1999) metonymization of black and white males and females, the bull is the "Supermasculine Menial," strong, brute, virile, alienated from the mind, with a penis but no phallus, yet who can get the phallus through the Ultrafeminine white woman. For Cleaver the "Ultrafeminine," embodied by the white woman, who is the paragon of beauty, is also frigid, weak, delicate, helpless, as well as a threat to the "Omnipotent Administrator" (white man) and the symbol of freedom for the Supermasculine Menial (black man). While wiping the black woman, or the "Subfeminine Amazon," from view, the cuckold scene may enable the Supermasculine Menial to use the white woman herself as his own prosthesis (to penetrate male whiteness through female whiteness), yet once the scene is over the prosthesis is promptly dismantled and he is left with nothing but the feeling of someone else's mission accomplished. Dyer had something to say about this mission:

> It [the naked body] may betray the relative similarity of male and female, white and non-white bodies, undo the remorseless insistences on difference and concomitant power carried by clothes and grooming. The exposed white male body is liable to pose the legitimacy of white male power: why should people who look like that—so unimpressive, so like others—have so much power? (Dyer, 1997, p. 146)

The introduction of the bull in the pornographic image, which features the purveyors of heterosexuality in their naked exposure, keeps the illusions of gender difference fundamental to the heterosexual logic from floundering: the blackness of the bull's skin will not confound itself with the whiteness of the woman's, or the also-maleness of the cuckold. We read contrast as difference. The interraciality of cuckoldry, then, serves as an immediate visually emphatic cue that suggests all of the drama and necessarily unbalanced power relations of heterosexual desire.

In many ways, therefore, cuckoldry has become such an alluring pornographic proposition for digital subjects because so many previously unflappable borders have broken down in the perverse relationship they can so easily have with their gadgets. In a digitally enabled context of perversion, in which so much feels possible—even the avoidance of material consequence (many videos fetishize unprotected sex with strangers as if these were either fictional or invincible bodies, much like the phallus)—the black body is called in to keep heterosexuality in the digital age from going down the wayside completely. The bull appears as the stanchion keeping heterosexuality's organization of bodies from shattering.

If the built white body is, as in Dyer's (1997) reading of muscle man films, not the body that white men are born with, but the body made possible due to his natural superiority, the body of the bull is born this way in its relationship to the Symbolic. If the white man's muscles are a product of thought, application, and planning, the bull's body is his blessing and his curse from the beginning. Whatever benefits the bull may reap from his labor are simply concessions compared to the overall work he accomplishes *for* the cuckold (in and out of the bedroom), which promptly belittles the bull's pleasure as incidental, short-lived, and counter-productive. This is a train that loops around its very own axis.

References

biguy60123. (n.d.). *Cuckold, creampie cleanup* [Video file]. Retrieved February 19, 2012, from http://www.xtube.com/watch.php?v=BHm93-G146-

Bruns, A. (2008). *Blogs, Wikipedia, Second Life and beyond: From production to produsage.* New York, NY: Peter Lang.

Cleaver. E. (1999). The primeval mitosis. In *Soul on ice* (pp. 205–220). New York, NY: Delta.

Colomina, B. (2007). *Domesticity at war.* Cambridge, MA: MIT Press.

CuckoldFantasies.com. (2012). Retrieved February 19, 2012, from http://www.fantasies.com/cuckolds_and_bulls.html

Cuckold Space. (n.d.). Retrieved February 19, 2012, from http://www.cuckoldspace.net/photo/wife-writing

Derrida, J. (1993). *Aporias* (T. Dutoit, Trans.). Palo Alto, CA: Stanford University Press.

Dyer, R. (1997). The white man's muscles. In *White: Essays on race and culture* (pp. 145–183). London, England: Routledge.

Fanon, F. (1967). *Black skin, white masks.* New York, NY: Grove Press.

Foster, H. (2006). *Prosthetic gods.* Cambridge, MA: MIT Press.

HKHORNY (n.d.). *Sexy Asian woman cuckolding with black guy* [Video file]. Retrieved February 19, 2012, from http://www.xtube.com/watch.php?v=q3gKc-S343-

Israël, L. (1996). *La jouissance de l'hystérique.* Paris, France: Éditions Arcanes.

Lacan, J. (2007) The signification of the phallus. In *Écrits* (B. Fink, Trans.). (pp. 575–584). New York, NY: W.W. Norton.

LT42011 (n.d.). *Cuckold fucked in motel* [Video file]. Retrieved February 19, 2012, from http://www.xtube.com/watch.php?v=YZE3D_S223_

Melcuck. (n.d.). *Fucking her BBC raw for the first time* [Video file]. Retrieved February 19, 2012, from http://www.xtube.com/watch.php?v= e7kNw_S531_

Melcuck (n.d.). *Cuckold double cream pie* [Video file]. Retrieved February 1, 2012, from http.//www.xtube.com/watch.php?v=ybQh9_S626_

Preciado, B. (2000). *Manifeste contra-sexuel*. Paris, France: Balland.

Rubin, G. (2004). The traffic in women. In J. Rivkin & M. Ryan (Eds.), *Literary theory: An anthology* (2nd ed.) (pp. 770–794). Malden, MA: Wiley-Blackwell.

SexyCassie. (2011). *Wife cheating with stranger in motel* [Video file]. Retrieved February 15, 2012, from http://xhamster.com/movies/ 526256/wife_cheating_with_stranger_in_motel.html

Tara_Tainton (n.d.). *Knocking my small-dicked cuckold boyfriend down– Part 3* [Video file]. Retrieved February 15, 2012, from http://www. xtube.com/amateur_channels/play.php?preview_id=vjL2VFAv___&type =preview

wimp2black. (n.d.). *Cuckold 1* [Video file]. Retrieved February 19, 2012, from http://www.xtube.com/watch.php?v=xrsiV_S418_

zen_sun84 (n.d.). ****** *Cuckold* [Video file]. Retrieved February 15, 2012, from http://www.xtube.com/watch.php?v=c5Vap_S228_

CHAPTER TEN

"An Internet Family": Online Communication During Childhood Cancer

Catherine McGeehin Heilferty

I was a young nurse at a metropolitan children's hospital working with infants who had graduated from the neonatal intensive care unit when, because of staffing needs, I floated to the oncology unit for the day. In that one day I experienced the full spectrum of what I had imagined nursing to be in its essence: working with heart and head, caring through science. Working with children with cancer and their families over the next 15 years, in the hospital, the outpatient setting, and in home care, permitted an appreciation of the struggles and gifts experienced. My view from sitting with families, listening to their fears, pain, accomplishments, and joys inspired many questions about how their communication styles and formats helped or hindered their quality of life. At the most intimate level, I began to appreciate what mattered most to families facing childhood life-threatening illness and the potential or actual loss of a child.

Years later, as a doctoral student, I began to develop a theory of online communication during illness. I read blogs created by parents attempting to manage the events, feelings, and decisions during life with childhood cancer. I returned every day to one blog in particular during an especially emotionally charged time in the family's life and remember thinking that the blogger was writing for a purpose, and that her motivations might be shared by other parents blogging about children with cancer—and that scientific inquiry should expand healthcare into the digital realm.

Cure rates for certain types of childhood cancers have improved; more families experience childhood cancer as a chronic illness. Indeed, a new illness trajectory has emerged that brings to the experience prolonged and complicated uncertainty and stress (Alderfer, Cnaan, Annunziato, & Kazak,

2005; Mishel, 1983). Research to date on the experiences of parents living with childhood cancer has focused on parental decision making, hope, uncertainty, and stress, as well as child, parent, and family quality of life or coping (Hinds, Burghen, Haase, & Phillips, 2004; Mishel, 1983; Stewart, Pyke-Grimm, & Kelly, 2005). Several studies have found posttraumatic stress symptoms in parents of children being treated for and surviving cancer, suggesting a need for greater attention to and prevention of the effects of the perceived stressors (Alderfer et al., 2005; Norberg, Lindblad, & Boman, 2005; Pöder, Ljungman, & von Essen, 2008). A growing body of pediatric oncology research centers on posttraumatic *growth* and highlights the positive outcomes that can result from negative life events (Barakat, Alderfer, & Kazak, 2006). Instruments have been developed to study parents' and children's resilience and benefit-finding behaviors (Phipps, Long, & Ogden, 2007). These theoretical frames inspired exploration of new areas of the childhood cancer experience. I examined parents' use of online technologies for ways in which blogging might enhance clinical decision making, relieve burdens, help authors manage prolonged uncertainty, and improve quality of life.

In this chapter, I present a theory about online communication during childhood cancer. The theory began with a belief. As I read blogs created by parents of children with cancer, I came to believe that parents use the Internet in a distinct way to solve problems related to the child's illness and to improve the family's quality of life. I read stories replete with strategies for managing stress and uncertainty, decreasing isolation, preserving family and personal identity, and promoting satisfaction and confidence. I questioned whether these strategies were underappreciated in current healthcare practice, and searched for research on blogging during illness.

Online communication as a means of production and distribution—of expression, of queries, of connection—has led to recognition of virtual communities. Baym (1998), Jones (1995) and Rheingold (1993) explored the concepts of online identity and online relationships. These researchers found evidence for support of the idea that virtual communities are no less real than corporeal groups; that togetherness and support can have the same effects online as in person. Research into online communication emphasized understanding belonging, boundaries, and authenticity, and ultimately the translation of online connection into action (Jones, 1995; Sharf, 1997). In less than 20 years we have witnessed the evolution in patients' use of the Internet from information source to support mechanism (Orgad, 2005; Sanford, 2010), empowerment device, source of emancipation (Keren, 2004), and advocacy and cultural resistance tool (Gillett, 2007). Hints of the therapeutic potential of the Internet for young people living with a chronic illness were discovered by Malik and Coulson (2011), who found that messages contained many expressions of personal experiences and requests for information and advice. Facebook groups have become a popular tool for awareness-raising, fundraising, and support-seeking related to breast cancer (Bender, Jimenez-Maroquin, & Jadad, 2011).

Rains and Keating (2011) found that, for bloggers with limited support from in-person relationships, reader support was negatively associated with loneliness and positively associated with personal growth. Thorough examination of online communication during illness can add another dimension to the role of caregivers in the illness experience. Hartzler and Pratt (2011) demonstrated that patients possess significantly different but equally valuable expertise than do clinicians, with the latter presenting information on the healthcare delivery system, biomedical research, and health professionals' work, and the former offering guidance on managing illness from day to day and advice about handling responsibilities and activities associated with family, friends, work, and the home.

What Are These Stories? Illness Blogs Defined

Early analyses of online communication on the topic of illness examined listservs, chat rooms, online support groups and other media of conversational modes. Exploration of breast cancer personal Web pages (the precursor to blogs) revealed information about the uses of the Internet and online writing, as well as the meaning assigned to the concepts of sickness, healing, recovery, and survival, among others. The new medium was used to find and share medical information and experiences, create peer support networks, construct bodies and selves in virtual space, and as a feminine political voice on the topic of illness within the context of society and culture (Pitts, 2004). Orgad (2005) explored online discussions and stories by breast cancer patients and family members for themes related to the meanings assigned to concepts such as support, stress, coping, and adaptation. McNamara (2007) identified self-expression, writer-reader interactivity, and control over the narrative as unique structural features of blogging breast cancer. Most broadly described as online journals or diaries, blogs are easily and regularly modified Web pages, posted in reverse chronological sequence which can be commented upon by readers, and may include links to other Web sites or blogs and more recently include images, either still or moving (Blood, 2002; Bruns, 2008; Herring, Scheidt, Wright, & Bonus, 2005).

The first step in theory development after a comprehensive review of the literature is to define in as clear and simple terms as possible the concept at the center of exploration (Walker & Avant, 2005). My analysis of the concept of illness blogs laid the foundation for the narrative analysis of parent blogs of childhood cancer, presented here. As the first step in concept analysis, I developed theoretical and operational definitions of illness blogs. Second, I identified essential attributes, concept antecedents, and consequences. Finally, I described the relationships between producers and users, or authors and audience (Heilferty, 2009).

An *illness blog* is the online expression of the narrative of illness. It permits an exchange of information, ideas, and emotions between affected and unaffected persons; family and strangers join in the story of an illness experience to help the affected manage and cope with the stress, uncertainty, and identity changes that accompany illness. Illness blogging is unsolicited, first-person writing by individuals on the topic of illness, published online for others to read and react to, which includes opinion, descriptions of experiences, reports of events and feelings, and perhaps the exchange of treatment information. Illness blogs involve a high degree of author-reader interactivity, possess a beginning and middle, but not necessarily an ending, and can either be publicly accessible or password-protected. They are published for the audience in real time and are created in the context of the illness. One defining characteristic of illness blogs is the blogger's unique "from me to the world" perspective. Whereas listservs and other chat formats rely on user-to-user interaction to sustain dialogue, bloggers create monologues that do not require but often welcome comment from readers (Heilferty, 2009). Ultimately, the analysis of the concept of illness blogs provided theoretical boundaries and inspired the first set of research questions into online communication during the childhood cancer experience.

Research Questions, Sample, and Analysis

This chapter reports on a part of a larger project answering three research questions. Due to space limitations, only two questions will be considered here: (1) What themes are evident in the illness narratives contained in blogs created by parents of children with cancer? and (2) What is the influence of author-reader interactivity expressed in illness blogs about the experience of parenting a child through cancer? For more information about the third research question, considering the life stories evident in these blogs, see Heilferty (2011b).

The purposive sample of parent blogs included publicly accessible, English language blogs created and authored by parents that contained descriptions of life with a child who had undergone treatment for acute lymphocytic leukemia (ALL) or neuroblastoma in the last five years, or who was currently undergoing treatment for these types of cancer. Blogs of interest were found using the search terms "diagnosed with leukemia" and "diagnosed with neuroblastoma" in the search features of CarePages, Caring Bridge, and Google. Online writing on these two types of cancer was expected to illustrate similarities and differences in family experiences of childhood cancer because childhood leukemia entails a relatively short, intense trajectory with high survival rates, whereas neuroblastoma involves exacerbations and recurrences over many years and has a much poorer long term survivorship.

Parents of nine children with neuroblastoma and five children with ALL consented to participate in the research by replying to an email solicitation. Of the 14 blogs in this study, 11 of the main authors were women and all authors were over 30 years old, consistent with recent reports on demographics of cancer bloggers (Kim & Chung, 2007).

Blog text totaling 15,479 pages, with photographs deleted, was copied and pasted into a Microsoft Word document, which was saved on a secure personal-use laptop. Names included in the analysis were excluded or changed to enhance privacy protection. The text of entire blogs, including comments by readers, was analyzed. Blogs were analyzed as individual units of illness experience expression and in relation to one another to identify thematic and linguistic similarities, consistent with Riessman's (2008) method of thematic narrative analysis. In analyzing blogs using Riessman's approach, narratives were organized into categories of themes, and analyzed until a stable set of concepts became evident.

Ethical Considerations

Before beginning research into lives complicated by illness, it was imperative to assess and explore potential ethical dilemmas that were inherent in or might result from the work (Heilferty, 2011a). Institutional Review Board approval was obtained before study initiation. Only blogs created by parents who gave informed consent to research were included in the analysis. Once a blog was selected for potential inclusion, individuals were contacted by email for consent to have their blog included in the analysis from the date of consent back chronologically to the first blog entry date. In research involving children, even indirectly, as in this case, the obligation to solicit assent to the research from the child must be considered.

Because the blogs were publicly accessible, with the parental value of strict privacy presumed less than consequential, and because content would be de-identified, assent from the child was not solicited.

Each research question required the use of a slightly different but essential perspective, reminiscent of the views during a long mountain hike. From below and far away, the mountain is taken in panorama. Study of the life stories evident in the blogs as reported by Heilferty (2011b) represented this panoramic view. The research questions addressed herein turned to the closer view on the hike, in which sights and sounds are sensed intimately, as in the feel of birch bark or a grasshopper's click. The themes of illness blogs were found through thematic analysis of individual blog entries as stand-alone narratives. Another perspective on a hike involves appreciating the interactivity between the walker and the woods, the influence of one on another. The blogs were examined for signs of influence between author and reader to answer the remaining question.

What's in the Stories? Illness Blog Themes

The answer to the research question, "What themes are evident in the illness narratives contained in these blogs?" resulted in a set of concepts being balanced by parents during childhood cancer diagnosis, treatment, and off-treatment phases. Iterative reading of the blogs, each time highlighting more expansive thematic elements in the narratives, led to clarity. It was during this phase of reading and rereading, pulling blog entries out of the context of the larger narrative, that the longing for balance became apparent. The small stories selected as significant were held together by the common thematic drive by the bloggers toward balance of negative forces with positive.

Balance and Ballast: Themes of the Illness Blogs

Five elements of the childhood cancer experience were found to be held in balance by bloggers. Effort toward balance between uncertainty and its management, stress and its management, change and constants, burdens and gifts, and private and public lives were found in each parent blog. Reflection on these thematic categories led to their organization into counterbalancing concepts within the overarching theme of balance.

Uncertainty and Uncertainty Management. From the time parents learned of the diagnosis, the blogs included narratives of the frustration and anxiety accompanying the many levels of uncertainty related to the illness, its treatment, and the changes it brought. For many parents, this uncertainty became the filter through which they viewed every day events and processed the experience. Uncertainty was described as torture. Metaphors illustrated feelings of powerlessness, chaos, and distress. The illness trajectory was a roller coaster, the disease a monster, a beast, the "thing under the bed." Parents and children engaged in a battle with uncertain outcomes. Bad news came out of the blue. Good news was suspect, greeted with alternating doses of thankfulness, skepticism and wariness of a short half-life. Efforts toward uncertainty management such as keeping records of events, information-seeking, and an active faith life, among many others, were found throughout all phases of treatment.

> Waiting is a funny thing. One moment you focus on all of the positives and the next minute you are considering all of the irrational conspiracy theories of why you don't know anything yet.... Did they find something and are waiting to tell us anything pending further review? Are they waiting to tell us because they want to give us more time to have happiness before they unleash the news from hell? Is it so bad or so odd that she is waiting to tell us until she has consulted with others and developed a plan? Is she waiting to tell us when we will be there in person.... My mind then jumps right back to the rational side as I evaluate all of my scary thoughts.... how can I sit here and debate conspiracy theories in the back of my head? It is this stupid evil disease. It is the fear. It is the unknown.

Entries described specific strategies that showed attempts to manage the uncertain future:

> We will keep on hitting the "play" button every day, trying not to "rewind" too far into the past, trying not to "fast forward" too far ahead into the future. For now, I feel deep peace in the midst of great sorrow. I am in a quiet, comfortable hotel room with my precious girl three feet away, happily and contentedly reading one of her beloved books.

Expressions of stress and anxiety in blog entries during periods of uncertainty were common. Blog writing was a tool that enhanced feelings of control over emotions. The narratives in this study highlighted uncertainty as changing in quality and intensity through the three phases of life with childhood cancer. In addition, differences were noted in the narratives of uncertainty written by families living with ALL and with neuroblastoma.

Blog entries frequently included references to or quotations from the Bible and other texts. These seemed to play a role in uncertainty management and to mitigate the sense of powerlessness when parents felt unable to provide concrete expressions of help to the child. Prayers and requests for prayers represented action. Written expressions of faith were particularly common at times of critical decision making. One parent used Biblical quotations as a framework for every post, seemingly to inspire, to guide her writing, and to frame perceptions of the events of the day.

Stress and Stress Management. Authors repeatedly described the blogs as a place to express feeling overwhelmed, powerless, frustrated, and sad. Indeed, blogs seemed the one place to express negative emotions and attitudes that authors described as too painful, too complex, or too isolating to relay in person even to close relatives and friends.

> Since we have been home it is so easy to forget about the cancer world. Yes we have [Jeremy's] daily care and the constant reminders when looking at his scars.... I am struggling with this whole "monster under the bed" feeling that at any moment it will come crawling out and attack us once again. The parents who are in the monitoring phase call it "scanxiety"...all of the anxiety and fear that is wrapped up in the scanning and waiting phase.

In a single blog entry, writing often shifted from a report or assessment of stressful encounters to a narrative of personal growth resulting from the perceived trauma of the illness and its treatment.

> Yes, she was diagnosed on our wedding anniversary! Eight years married and a little over ten together. And let me say, that the fact we are together today after the stress of the last two years, means we'll be together forever!

For some, having a plan in place for potential negative outcomes helped to diminish stress levels.

> So, having heard of this back up plan, we are much more at ease with the knowledge that we may be at or near the end of our run with this current chemo. Having a backup plan makes getting through the day much, much easier.

At times, blog entries expressed a hope for purpose in the sharing of so much information with readers.

> I feel as if I complain about the hospital/clinic days often. It's not my intention, but for those of you not on the front line and rather on the interior brigade, I hope it makes for good reading. I would feel great knowing that there is some value to the hell we endure. Additionally, as I've typed before, this is the place I purge. If I can purge...then I can release it. Besides, she needs to read this one day and know what a courageous child she was.

Change and Constants. Each blog contained references to a changed life after cancer and descriptions of a new normal. In fact, several phases of change representing several new normals were noted. Upon reflection, life during the diagnosis phase was often the period of highest stress and uncertainty, when the first new normal began, i.e., the end of life before cancer.

[W]e took her and Martin to a kid's movie. I was amazed at how different everything felt—Dave and I have discussed that life seems to be divided into "before diagnosis" and "after diagnosis." Before her diagnosis, I took her laughter for granted.... But sitting beside her in the theater, after having gone through two weeks of pain and uncertainty and fear together, each smile and giggle was a beautiful treasure. Her smiles before diagnosis were worth about a penny, they were plentiful and effortless. Her smiles after diagnosis are worth more than diamonds.

The second new normal seemed to emerge during treatment. Work and school obligations were met, albeit with interruptions for hospital and clinic visits. Knowledge about the illness accumulated and confidence in the ability to meet the child's emotional, medical, and physical needs grew. Parents expressed unhappiness and fear related to the illness experience and the associated uncertainty, but during this time bloggers also wrote of an awareness that some of the families' constants before cancer remained: abiding love, financial stability, the comfort of daily routines, faith, and family traditions.

It is such a strange thing this new normalcy that we have. I realize most people would think it would be great and it is however it is also unsettling.... I guess this is the NEW normal you hear talked about around clinic and the doctors. When things were just completely cancer you never thought to compare it to what it should be, you just knew it was all wrong. Well now we are trying to strike a balance between cancer and normal life.... One thing cancer has taught us is that most things aren't that big of a deal in comparison to cancer. It definitely has put things in perspective, however normal life consists of a lot of little issues that we can't totally blow off. Trying to strike a balance is a challenge for us.

Changes in personal and family identity were evident in many blog entries.

I've been pondering alot recently about life and how different mine is now. I am the same person but I'm not. There is something very different about me and it's not visible until I tell someone my daughter had cancer. The look on their faces changes instantly.... Sometimes I leave people speechless and I don't mean to. Other times, people will ask about it and once they hear it all, I don't really hear from them again. I have some friends and family, but most of the time I feel so isolated...completely alone. The friends I do have I rarely talk to and I only talk to a couple people in my immediate family too.... I have wonderful friends from all over the country that we have met through Sloan. They are the best friends because they "get it".... I talk to people about regular every day things, but it feels awkward to me. I don't and never will look at things in life the same way again.... I don't think I will ever lose that isolated feeling. I have a scar on my soul that has forever changed me.

Some reference to the positive effect of living through the experience was found in each blog. This was especially true during the off-treatment phase or after a child had died. Parents expressed surprise and pride in having learned that despite the obstacle that childhood cancer represented, the identity change that resulted was interpreted as positive. The individual and family identity changed but bloggers learned lessons they otherwise might not have. Personal strength and an appreciation for life were frequently mentioned enrichments to life after cancer.

Burdens and Gifts. The burdens expressed in the blog entries were many and varied. The principal burden, the most difficult to bear, was a sense of loss. Parents expressed feelings of sadness at the loss of normal life, separation, and everything from hair to missed celebrations to the loss experienced with a death of a child. Here a father described the influence of the loss of control.

> Once again, I see many themes that run true to several of the families in this fight....
> On a daily level, this illness can dictate when you laugh or cry, eat or go hungry,
> sleep or stay awake, work or stay home, watch tv or watch a monitor that is attached
> to your child, spend time with family or spend time with doctors and nurses....
> when I look back over the past 2.5 years, this disease has really dictated every move
> we've made. You try to retain control over your life and the lifestyle that you've
> provided for your family, but when sickness knocks on the door, you can't ignore
> it.... Try to plan a night out with friends, maybe for just a few hours, and see if a
> fever doesn't pop up out of nowhere and have you driving to the hospital for four or
> five unexpected days.... Yes, we function.... But...no matter how hard you try, it
> finds a way to control every aspect of your life. You may think you have control,
> but you have no more control than the surfer that waits for the wave.... they can't
> conjure one out of thin air, anymore than we can create a lull in the action to give
> ourselves a break.

Examples of surplus suffering, i.e., the imposition of discomfort, pain, or stress in addition to the predictable burden of the illness and its treatment, were found in every blog. Nearly every blog contained writing on the burdens of expense and separation associated with the travel required to receive care at distant centers where experts in these rare illnesses could be found.

> One woman we talked to said her son was diagnosed five years ago and has just
> relapsed for the second time. She said, "It never gets easier. You can never rest.
> Even if your child is declared N.E.D. (no evidence of disease) you don't relax." She
> very aptly summed up the other pressures of a family experiencing the disease—the
> financial strains, the marital strains and the pressures on the other children.

Parents also expressed feeling burdened with having to take on the role of educator when staff members were unfamiliar with equipment or procedures. During hospital stays, witnessing perceived incompetence and the suffering of other children and families living with cancer were intensely difficult to bear. For one father, the aggravation of communication with health insurers rivaled the burden he experienced in his child's death. For the rest of the bloggers, communicating with insurers took time and energy away from life with the child at a time of great need.

The discussion of gifts supports the notion that writing in general, and blogging in particular, engender a sense of community. Authors noted acquiring gifts of personal growth; an awareness of strength; and an appreciation for life, for family, and for what's really important during the treatment and off-treatment phases.

> One of the questions that many people ask when faced with a crisis such is this is: "Why did this happen?" Now I don't know why, but I haven't asked myself this question and I guess it is time to face this demon....it has never occurred to me....I guess it might be because of the way that I look at life.... I believe I learned to deal with life's little hurdles when my dad passed away.... I spend my time looking for the good, looking for that special nugget of information that I should learn.... I care more. I love more. And, most importantly, I appreciate what I have even more. We have a long road ahead and lots of learning to do. So for the time being, it really doesn't matter why this happened, but rather, what are we going to take from it?

Private and Public Lives. Parents' identities as private persons and public entities oscillated as time passed. The community of readers grew from family members and close friends to a broader audience interested in events, in the narrative, and in providing support by writing and by taking action. Prayer, companionship, and raising both funds and awareness were considered acts of support from readers. At times, parents wrote of feeling conflicted about how much information to share and when. Ultimately, the benefits of taking on something of a public persona appeared to outweigh side effects such as reader misinterpretation, negative feedback, and reader suggestions perceived by bloggers as unhelpful, even hurtful.

> I feel a little overwhelmed lately be the responsibility of sharing my family's life with everyone who reads our blog. Certainly we are in awe of the number of people who have read about Jeremy's story and we are humbled by how many are praying! I guess I just want everyone to know that in my imperfection I might now always write clearly, might now always explain things in the right way, and might not always do things the way other's think I should.... I said this when I first started writing and I'll say it again, I'm blogging to tell the story of my precious son Jeremy who has cancer. I've since realized that by sharing the story of this difficult journey, I'm also sharing our faith. It is my prayer that I'm bringing glory to God through this blog, despite my imperfections.

The blurring of public versus private life when blogging about illness remained despite some parents' use of password protection. Two blogs were registered in the public domain. The remainder used what is now referred to as light password protection, (registration with Web sites using only self-created usernames and passwords). Parents understood the degree to which their privacy was protected. Although no violations of privacy were mentioned, parents described violations of trust, such as when a hospital executive referred to information from a blog during a conversation with parents about family dissatisfaction with systems issues. Universal standards of public versus private content and acceptable behavior online remain undefined, but to these bloggers and their regular readers, codes of conduct were very clear (Heilferty, 2011a; Sievers, 2006).

Ballast and Co-Creation: Author-Reader Interactivity in Illness Blogs

The second research question, "What is the influence of the author-reader interactivity expressed in illness blogs about the experience of parenting a child through cancer?" explored yet another layer of blog purpose and effect. A defining feature of illness blogs, interactivity was the nexus at which author and reader connected in parents' search for balance.

One reader described the relationship between blogger and readers as an Internet family:

> I've been here…sometimes posting in the guestbook - a lot of time just reading.... I pray that you have numerous friends, family, and internet friends to be by your sides as this journey ends. Please know that we as an internet family are out here thinking of you during all this. May [your child] sleep the sleep of peace and know that he has been much loved in his short lifetime and that he has changed so many of our lives.

Co-creation of a broader narrative of childhood cancer resulted as bloggers and readers shaped not just the dialogue but the experience itself as well as its perception.

The lasting effects of the interactivity were most apparent in the treatment phase, when readers had the greatest influence on the authors' subsequent writing. During the diagnostic phase, reporting details and events predominated. During the off-treatment phase, parents were either deep in the isolation surrounding the loss of the child, or they were better able to maintain the balance inherent in survivorship. However, during the treatment phase, comments from readers provided ballast that lifted spirits, affirmed feelings, and helped parents solve problems.

As expected, comments from readers influenced the bloggers' writing, but the degree to which the comments influenced the illness experience itself was surprising. Bloggers expressed the need to hear from readers regularly, even in times of stability. The blogs truly became works of co-creation, influencing the bloggers' perception of the negative forces and inspiring an emphasis on positive thinking.

What's to Be Done With Stories? Contextualizing and Further Developing the Theory

For the family members in this study, blogging was a forum for the exchange of ideas, experiences, and knowledge. Parents reported psychosocial, physical, spiritual, developmental and financial stresses as well as management strategies to address these, confirming the need for varied modes of support from healthcare providers (IOM, 2007). Parents' writing included evidence of treatment efficacy and adverse effects; barriers to compliance with therapy; and their understanding of the illness, its treatment, the healthcare system, and the education they received about all three. They advocated for change, suggested areas for improvement to the healthcare system, and lobbied for expansion of treatment research initiatives both locally and nationally. Many parents in this study expressed a desire for greater control over the flow of health information, especially about result reporting. Differences in parent and physician expectations regarding communication found in the blogs suggest a need for creation of new, more efficient means of discussing care.

These findings are consistent with recent Internet research. As in McNamara's (2007) work, the parent blogs appeared to relieve suffering by helping individuals maintain close relationships despite separation and by aiding in the formation of new relationships, regardless of temporal or geographic proximity (McNamara, 2007). My research supports the idea of Baym (1998) and Rheingold (1993) that online and offline identities and relationships are equally meaningful and full of potential. Evidence was found in the analysis of comments consistent with Orgad's (2005) finding that communication in breast cancer listservs aids meaning-making. Clearly, one of the effects of readers on bloggers in this study was the enhancement of bloggers' ability to make meaning of the illness experience and to cope with stress and uncertainty—providing ballast for parents feeling off-balance.

Bloggers' descriptions of unwelcome thoughts and flashbacks of stressful events and nightmares support Kazak, et al.'s (2004) study of PTSD in parents of childhood cancer survivors and research by Pöder, Ljungman, and von Essen (2008), who found parent reports of symptoms of PTSD within one week of childhood cancer diagnosis and reports of PTSD symptoms two months and four months later. Evidence of personal growth found in the blogs was consistent with work done by Barakat, Alderfer, and Kazak (2006) and Phipps, Long, and Ogden's (2007) benefit-finding behaviors.

Narratives of both surplus suffering and unexpected gifts were found in the blogs. Significant burdens identified by bloggers, in addition to those associated with the illness or its treatment, included financial or work-related stresses, additional expenses and parental feelings of isolation during travel for treatment, consistent with the findings of Clarke and Fletcher (2004). The gifts reported by parents included the ability to maintain the capacity to have plans and desires, to hold on to a sense of themselves even with somewhat limited control over their lives; and a renewed clarity about life's important elements.

Future research is needed to further refine this developing theory, and to inform healthcare practitioners' efforts to incorporate new communications media into the overall provision of care. Participatory medicine, gaining popularity as a new model for healthcare, might include expansion of the electronic health record to incorporate space for parents to report symptoms, problems or questions (Society for Participatory Medicine, 2011). Some healthcare systems refer to these as personal health records (PHRs) and use them to email patients, allowing patient self-scheduling and results reporting. To date, theorizing and planning in health informatics has been driven by professionals' perceptions of patients' online activity as solely information seeking. Production of and contribution to electronic medical records has been one-sided, with patients as end-users having little control over content, context, and distribution (Winkelman, Leonard, & Rossos, 2005). Nursing informatics has charted a course through 2018 to expand users of interest to include interdisciplinary researchers, guide the reengineering of nursing practice, harness new technologies to empower patients and their caregivers for collaborative knowledge development, and to facilitate the development of middle-range nursing informatics theories, among other priorities (Bakken, Stone, & Larson, 2008).

Thus far, work toward a theory of online communication during illness explains the influence of Internet use on the illness experience and the meaning of author-reader relationships. I believe online communication and relationships between authors and readers improve quality of life during illness, and may improve health outcomes.

New knowledge about information gathering, supportive relationships, and perceived quality of life can expand the limited understanding of the impact of parents' use of the Internet as a means of communication during illness. Assessment of the efficacy of online community interaction in decreasing uncertainty, stress, and suffering could be the foundation for interventions aimed at enhancing parent-provider relationships, promoting personal and familial growth, and improving care quality. The ballast provided by the members of the virtual community of childhood cancer blog readers merits further study. Of special interest is the description of the relationship as that of "family" by some bloggers and readers, a tag that goes a step beyond the present understanding of virtual communities.

Although illness blog analysis has grown from nonexistent to a small-scale body of evidence, less is known about parents' use of other forms of online communication. Social media offer the same immediacy of storytelling, albeit abbreviated. More needs to be learned about the use and effect of these platforms of expression to more fully incorporate the family in the overall plan for care. I intend to conduct a grounded theory study of parents' use of the Internet during childhood cancer, and to interview parents regarding social media use during illness to learn more about its effect on the experience. I expect that this will lead to new ways healthcare providers might facilitate parents' ability to acquire and use health information, manage illness, and improve family quality of life. The virtual community, the Internet *family*, clearly offers support and guidance to parents. Future study will explore the virtual family's capacity to empower parents to become agents for personal, family, and communal growth. As one blog author in my sample put it:

> All of you wonderful people that I have met [online] have become like a family to me. You understand my pain and my joy.... I always think of how lucky we were...being there for other families who aren't as lucky as us has been my quest.

References

Alderfer, M., Cnaan, A., Annunziato, R., & Kazak, A. (2005). Patterns of posttraumatic stress symptoms in parents of childhood cancer survivors. *Journal of Family Psychology, 19*(3), 430–440.

Bakken, S., Stone, P. W., & Larson, E. L. (2008). A nursing informatics research agenda for 2008–18: Contextual influences and key components. *Nursing Outlook, 56*(5), 206–214, e3.

Barakat, L. P., Alderfer, M. A., & Kazak, A. E. (2006). Posttraumatic growth in adolescent survivors of cancer and their mothers and fathers. *Journal of Pediatric Psychology, 31*(4), 413–419.

Baym, N. (1998). The emergence of on-line community. In S. Jones, (Ed.), *Cybersociety 2.0: Revisiting computer-mediated communication and community* (pp. 35–68). Thousand Oaks, CA: Sage.

Bender, J. L., Jimenez-Maroquin, M. C., & Jadad, A. R. (2011). Seeking support on Facebook: A content analysis of breast cancer groups. *Journal of Medical Internet Research, 13*(1). Retrieved January 12, 2102, from http://www.jmir.org/2011/1/e16/

Blood, R. (2002). *We've got blog: How weblogs are changing our culture.* Cambridge, MA: Perseus.

Bruns, A. (2008). *Blogs, Wikipedia, Second Life, and beyond: From production to produsage.* New York, NY: Peter Lang.

Clarke, J. N., & Fletcher, P. C. (2004). Parents as advocates: Stories of surplus suffering when a child is diagnosed and treated for cancer. *Social Work in Health Care, 39*(1/2), 107–127.

Gillett, J. (2007). Internet Web logs as cultural resistance: A study of the SARS Arts Project. *Journal of Communication Inquiry, 31*(1), 28–43.

Hartzler, A., & Pratt, W. (2011). Managing the personal side of health: How patient expertise differs from the expertise of clinicians. *Journal of Medical Internet Research, 13*(3). Retrieved January 12, 2012, from http://www.jmir.org/2011/3/e62/

Heilferty, C. M. (2009). Toward a theory of online communication in illness: Concept analysis of illness blogs. *Journal of Advanced Nursing, 65*(7), 1539–1547.

Heilferty, C. M. (2011a). Ethical considerations in the study of online illness narratives: A qualitative review. *Journal of Advanced Nursing, 67*(5), 924–931.

Heilferty, C. M. (2011b). *The balance we seek: A sequential narrative analysis of childhood cancer blogs* (Doctoral dissertation). Retrieved from ProQuest Digital Dissertations. (AAT 879040861)

Herring, S. C., Scheidt, L. A., Wright, E., & Bonus, S. (2005). Weblogs as a bridging genre. *Information Technology & People, 18*(2), 142–171.

Hinds, P. S., Burghen, E. A., Haase, J. E., & Phillips, C. R. (2004). Advances in defining, conceptualizing, and measuring quality of life in pediatric patients with cancer. *Oncology Nursing Forum, 33*(Suppl. 1), 23–29.

Institute of Medicine. (2007). *Cancer care for the whole patient.* Retrieved March 18, 2008, from http://www.iom.edu/CMS/3809/34252/47228.aspx

Jones, S. (1995). Understanding community in the information age. In S. Jones (Ed.), *Cybersociety* (pp.10–35). Thousand Oaks, CA: Sage.

Kazak, A. E., Alderfer, M., Rourke, M. T., Simms, S., Streisand, R., & Grossman, J. R. (2004). Posttraumatic stress disorder (PTSD) and posttraumatic stress symptoms (PTSS) in families of adolescent childhood cancer survivors. *Journal of Pediatric Psychology, 29*(3), 211–219.

Keren, M. (2004). Blogging and the politics of melancholy. *Canadian Journal of Communication, 29,* 5–23.

Kim, S., & Chung, D. S. (2007). Characteristics of cancer blog users. *Journal of the Medical Library Association, 95*(4), 445–450.

Malik, S., & Coulson, N. S. (2011). The therapeutic potential of the Internet: Exploring self-help processes in an Internet forum for young people with inflammatory bowel disease. *Gastroenterology Nursing, 34*(6), 439–448.

McNamara, K. R. (2007). *Blogging breast cancer: Language and subjectivity in women's online illness narratives* (Doctoral thesis). Retrieved June 28, 2007, from http://www3.georgetown.edu/grad/cct/academics/theses/etd_klr25.pdf

Mishel, M. H. (1983). Parents' perception of uncertainty concerning their hospitalized child. *Nursing Research, 32*(6), 324–330.

Norberg, A. L., Lindblad, F., & Boman, K. K. (2005). Parental traumatic stress during and after paediatric cancer treatment. *Acta Oncologica, 44,* 382–388.

Orgad, S. (2005). *Storytelling online: Talking breast cancer on the Internet.* New York, NY: Peter Lang.

Phipps, S., Long, A. M., & Ogden, J. (2007). Benefit finding scale for children: Preliminary findings from a childhood cancer population. *Journal of Pediatric Psychology, 32*(10), 1264–1271.

Pitts, V. (2004). Illness and Internet empowerment: Writing and reading breast cancer in cyberspace. *Health: Interdisciplinary Journal of Health, Illness and Medicine, 8*(1), 33–59.

Pöder, U., Ljungman, G., & von Essen, L. (2008). Posttraumatic stress disorder among parents of children on cancer treatment: A longitudinal study. *Psycho-Oncology, 17*(5), 430–437.

Rains, S. A., & Keating, D. M. (2011). The social dimension of blogging about health: Health blogging, social support, and well-being. *Communication Monographs, 78*(4), 511–534.

Rheingold, H. (1993). *The virtual community.* Retrieved January 11, 2012, from http://www.rheingold.com/vc/book/index.html

Riessman, C. (2008). *Narrative methods for the human sciences.* Thousand Oaks, CA: Sage.

Sanford, A. A. (2010). "I can air my feelings instead of eating them": Blogging as social support in the morbidly obese. *Communication Studies, 61*(5), 567–584.

Sharf, B. F. (1997). Communicating breast cancer on-line: Support and empowerment on the Internet. *Women and Health, 26*(1), 65–84.

Sievers, L. (2006, May 19). *My cancer: Discussion guidelines* [Blog post]. Retrieved February 17, 2011, from http://www.npr.org/blogs/mycancer/2006/05/discussion_guidelines.html

Society for Participatory Medicine. (2011). *About us* [Blog post]. Retrieved February 20, 2011, from http://participatorymedicine.org

Stewart, J. L., Pyke-Grimm, K. A., & Kelly, K. P. (2005). Parental treatment decision making in pediatric oncology. *Seminars in Oncology Nursing, 21*(2), 89–97.

Walker, L., & Avant, K. (2005). *Strategies for theory construction in nursing* (4th ed.). Upper Saddle River, NJ: Pearson.

Winkelman, W. J., Leonard, K. J., & Rossos, P. G. (2005). Patient perceived usefulness of online electronic medical records. *Journal of the American Informatics Association, 12*, 306–314.

CHAPTER ELEVEN

The Identity, Content, Community (ICC) Model of Blog Participation: A Test and Modification

Brittney D. Lee and Lynne M. Webb

It is a picture we see every day in the United States. Mothers meet for lunch at a fast food restaurant, watch their children frolic in the bubble pit, and discuss preschool applications and potty training. Mothers sit on park benches watching their children play on monkey bars and swing sets as they converse with one another about childrearing philosophies and practices. At family reunions, young-adult cousins meet in the kitchen to discuss pediatricians and birthday parties for the next generation. Although face-to-face discussions of children and family continue, parallel conversations also occur among strangers online in chat rooms, on parenting sites, and on mommy blogs. This chapter examines the wildly popular phenomenon of mommy blogs and their growing online communities. We report a test of our Identity, Content, Community (ICC) model depicting such blog communities (Webb & Lee, 2011). The model describes how a mommy blogger's identity influences her blog content, and how content provides the basis for a sense of community among mommy bloggers—a community fostering close relationships and social support. The model also depicts symbiotic relationships among these factors, with these communities, in turn, influencing mommy bloggers' identity and content.

Blogging has been described as the offspring of personal Web pages and user-generated content (Turkle, 1995). Personal Web pages present original content, but are fairly static. In contrast, weblogs, or blogs, contain regularly changing content; new posts typically appear on a daily basis in reverse chronological order (Wei, 2009). Further, blogs usually allow readers to comment on each post, thus creating original, constantly changing content. Blogs can contain information about a wide range of topics from personal life to politics (Jost & Hipolit, 2006) and are often organized into genres based on content, including mommy blogs.

Blogs typically have three main characteristics (Droge, Stanko, & Pollitte, 2010): they provide original content in posts; they usually link to other blogs via a list of links (blogroll), thus creating networks or communities (the blogosphere); and most allow reader comments. From their earliest beginning to the present day, scholars have characterized blogs as a powerful medium of communication (Kline & Burstein, 2005; Rodzvilla, 2002; Rosenberg, 2009; Woods, 2005).

Blogs are often categorized as either filter blogs or personal blogs (Herring & Paolillo, 2006; Wei, 2009). A filter blog highlights previously produced information on a given topic, focusing on news and politics (Wei, 2009, p. 533). Political blogs, for example, often link to the Web sites of traditional media sources, such as newspapers. In turn, traditional media outlets often quote political filter blogs (Tucker, 2009).

In contrast, personal or diary blogs feature journal-like descriptions of daily life and informal photographs (Bruns, 2005; Jung, Vorderer, & Song, 2007). Although journal blogs may seem more personal than filter blogs (Stefanone & Jang, 2007), they are written for a mass and ambiguous audience that may include family, friends, acquaintances, and strangers (Kleman, 2007).

One popular genre of female personal blogs is called mommy blogs. A mommy blog, for the purposes of this study, is defined as a blog predominantly about family. Although mommy blogs have become the focus of increased scholarly attention (Camahort, 2006; Friedman & Calixte, 2009; Hammond, 2010; Lopez, 2009; Moravec, 2011), we could locate no previous empirical study of mommy blogger identities, content, or communities. Our ICC model, described fully in Webb and Lee (2011), articulates and explains the relationships among six basic elements: identity, motivation, content, community, support, and relationships (see Figure 11.1, a variation of the model presented in our prior work.). In this chapter, we report on a study examining the inter-relationships among three of the six elements (identity, content, community) discussed in detail below.

The ICC Model

For purposes of this study, we use the term *identity* to reference self-image or self-concept that manifests in behavior. Scholars have extended the concept of performing identity into the online realm (Baym, 2006; Mendelson &

Papacharissi, 2011), documenting that identity can be influenced by online associations, connections, community and relationships (Turkle, 1995). A blogger potentially enacts identity on her blog through her posts and, in turn, her identity can be influenced by the performance itself—creating a mutually influential relationship between the blog and the blogger's identity.

Mommy bloggers typically post about their personal experiences of motherhood and thus reveal their maternal identity—an identity often very different from the picture of motherhood presented in mainstream media. Lopez (2009) argued that mommy blogs don't focus on idealized depictions of maternal bliss, but rather "women who are frazzled by the demands of their newborn baby, who have no clue what to do when their child gets sick, who suffer from postpartum depression and whose hormones rage uncontrollably" (p. 732). By embracing the reality of motherhood as they experience it, mommy bloggers offer a more realistic portrayal of maternal identity.

For the purposes of this study, we defined *content* as the subject matter discussed on the blog. Although many people may think mommy bloggers discuss growth charts and dirty diapers, the blogs offer significantly richer content including major life issues such as adoption, charities, infertility, health, money, and work. Mommy bloggers see themselves as engaging in the political act of redefining motherhood and family care-giving in a public way on a national stage. For example, Alice Bradley of Finslippy (http://finslippy.typepad.com), described mommy blogging as a radical act, one redefining 21st century maternity (Lopez, 2009) and BlogHer (a blog hosting Web site) provided credibility and validity to mommy bloggers by holding a discussion session titled "MommyBlogging is a Radical Act!" at its 2006 conference (Camahort, 2006).

We use the term *community* to reference a sense of cohesiveness, commonality, and propinquity among a group. Blogs form online communities around a specific theme, idea, industry activity, or other subject of the blog (Davies, 2007; Droge et al., 2010; Gurak & Antonijevic, 2008; Hatzipanagos & Warburton, 2009; Humphrey, 2008; Kouper, 2010b). Community development is facilitated by two important blog characteristics: linking to readers' blogs and allowing or encouraging reader comments.

Castells (2004) has traced the rise of blog communities to the loss of real-life communities. For example, Shirky (2008) observed that Americans no longer have ready opportunities for a wide sense of community because they live in close quarters, have their own corner grocery store and neighborhood school district, and live in smaller subdivided communities. Enter blogs, where a "sense of community is developed through interactions with like-minded people" (Kaye, 2005, p. 76). Through blogging, authors invite audience members to discuss, share, and support one another (Lopez, 2009).

Mommy blogs were never developed to appeal to everyone but rather to fellow mothers in similar circumstances. Indeed, mommy bloggers typically write to other mothers, usually to fellow mommy bloggers, thus forming a specialized blogging community or "blogosphere" of blog authors who read and comment on each other's work (Webb & Lee, 2011). Further, individual mommy blogs tend to be quite focused in topic and certainly are not all the same or even similar. For instance, one mommy blogger community may be Midwestern farm wives and mothers who home school, whereas another might be moms who work on Wall Street and whose children attend private middle-schools. Communities form around a given blog because of common interests that vary with the age of children, geography, and activities. In sum, "blogs permit people in engage in social interactions, build connections, maintain conversations, share ideas, and collaborate with others" (Dumova, 2012, p. 250) in similar circumstances.

Such common interests and circumstances, however, require interaction for a sense of community to develop. Blanchard (2006) has argued that blogs become virtual communities when they offer interactivity through comments and regular posting by the authors. In a study of a community on LiveJournal, a blog-hosting site, users created a community by providing advice and talking freely to one another (Kouper, 2010a). Free-flowing communication on common topics allows for ideas to be shared. Warburton (2010) wrote that within online communities, female users reported feeling supported and not judged for their original ideas. The sharing of a common interest and original ideas contributes to community formation (Hine, 2004). Allowing readers to communicate easily, frequently, and regularly about specific topics contributes to blog popularity (Webb, Fields, Boupha, & Stell, 2012).

In addition to interaction, the blogger can facilitate a sense of community among readers and writers on the site by link-sharing (Li & Chignell, 2010). When a blogger links to 10 additional blogs on her blogroll, she builds connections to those 10 people. Bloggers often label their blogrolls with names that function as endorsements such as "Blogs I Read" or "Blogs I Recommend." As bloggers begin listing each other on their blogrolls, a community can form through these associations, and, conversely, blogrolls typically reveal bloggers' communities. Further, previous research has linked the number of hyperlinks on a blog to the number of hits the blog receives, a common indicator of popularity (Webb et al., 2012).

Testing the ICC Model

In previous work (Webb & Lee, 2011), we asserted that many of the characteristics of blogging discussed above appear interrelated. We begin our explanation of the model with the identity variable and its hypothesized relationships: As a woman is motivated to blog by her identity (Arrow #1), she chooses her blog content based on her motivations (2) and identity (3).

As she blogs, she may discover additional aspects of the self that affect how she thinks about herself (3). We acknowledge that women may have multiple motivations for blogging, in addition to expression of identity, such as financial incentives or the desire for a creative outlet. However, we assume that the vast majority of mommy bloggers blog as themselves (rather than an alternative identity) and thus their blog is, on its face, a straightforward expression of identity.

Many mothers choose to blog about their families, and become labeled mommy bloggers. The specialized content and the interactivity among fellow bloggers as they post on her blog creates a blogging community (4), where members give and receive support (5), and, in turn, form friendships (6), reinforcing a sense of community (7). The interactivity of these factors come full circle as a woman's identity is influenced by being a member of the blogging community (8). These factors (community, support, relationships) influence a blogger's identity (8) and in turn may serve as a stronger motivation to continue blogging (1), thus continuing the cycle of influence among identity, motivation, content, community, support, and relationships.

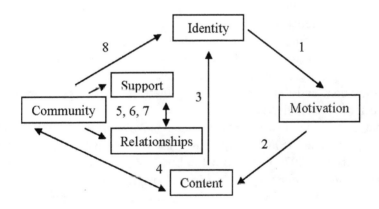

Figure 11.1. The ICC model of the mommy blogosphere. Based on Webb & Lee, 2011. Published with the permission of Cambridge Scholars Publishing.

Although the above-described variables and many of the relationships between them have been previously identified, our ICC model is the first to connect them and display their interrelationships in one cohesive model. In this, the first empirical test of the ICC model, we conducted a four-part, online survey (in counter-balanced orders to ameliorate order effects) of mommy bloggers to test the following hypotheses:

H1: Mommy bloggers' identity is related to their blog content (Arrow 3).

H2: Mommy bloggers' experience of community is related to their identity (Arrow 8).

H3: Mommy bloggers' content is related to their experience of community (Arrow 4).

After receiving approval from our university's Institutional Review Board, we recruited bloggers via an invitation to the online survey with a link posted for one day on Kelly's Korner Blog (kellyskornerblog.com). We encouraged bloggers who completed the survey to solicit additional recruits by linking to our survey on their own blogs and thus snowballing the sample.

Over 1,000 bloggers (n=1175) completed the online survey but fewer (n= 1157) met the inclusion criteria: self-reported as female residing in the United States who blogged primarily about her family. Our sample of 1,157 participants ranged in age from 20 to 66 (M=30.66, SD =5.62), and reported residing in every state except Montana, New Hampshire, and Vermont. The sample was predominantly well educated (77% at least some college), Christian (77%), and Caucasian (79%). Less than 1% of the sample self-reported in each of the ethnicity categories except Caucasian and nearly 20% of the sample failed to report ethnic identity.[1]

The survey contained four major instruments and a basic demographic questionnaire. We pre-tested the efficacy of the questionnaires and the administrative procedures with 54 participants recruited through online forums on thenest.com and thebump.com, using the responses to guide minor revisions prior to data collection.

Our four major instruments included Stake's (1994, 1998) Six-Factor Self-Concept Scale (SFCS), previously used with diverse female samples (Kwon & Rudd, 2007; Yanico & Chih Lu, 2000); it assesses six dimensions of *self-concept* or *identity*: gifted, moral, power, task accomplishment, vulnerability, and likeability. We conducted a component factor analysis employing varimax rotation with Kaiser Normalization across the SFCS scores; it yielded the anticipated six factors with Cronbach alphas of .81, .80, .84, .84, .83, and .84 respectively. To develop the six SFCS scores for each participant, item responses of 1 to 7 were summed across each factor for each participant, creating composite scores for the dimensions that ranged from 5 to 42 for each dimension.

Because tested and validated instruments could not be found to measure *content*, we designed an original questionnaire including questions adapted from the Princeton Research Survey Associates' Blogger Callback Survey (2006), Lenhart and Fox's (2006) Pew Internet and American Life Project blogging survey, and specific mommy blogging topics (Friedman & Calixte, 2009). The content measurement included sections identifying the three topics they most often wrote about, the label that best described their blog, what types of blogs they read, the types of blogs on which they commented, and how often they blogged about 25 specific topics. Because we assessed these variables to test predicted relationships in the model, we did not calculate demographic measures for them. However, specific topics discussed on blogs were analyzed, as discussed below.

A two-step cluster analysis yielded two distinct groups of mommy bloggers, distinguished by the content of their blogs. We labeled the larger group (88%) Family-Focused Bloggers and the smaller group (12%) Family-Plus Focused Bloggers, hereafter called simply Family and Family-Plus Bloggers. Both groups blogged about topics directly related to families (such as life within the family), but the Family-Plus bloggers posted more often about topics *outside* the family (including business, entertainment, health issues, money, news, philanthropies, photography, products, politics, political views, sports, technology, adopting, cooking, and home-schooling—which typically involved concerns outside the family such as purchasing curriculum, state certifications, and field trips with other home-school children).

Because tested and validated instruments could not be found to assess *experience of community*, we created such a questionnaire based on Blanchard (2006) as well as Haferkamp and Kramer (2008). The questionnaire included three sections assessing community participation, which asked whether participants read, wrote, and commented on blogs; how often they read, wrote, and commented on blogs; and about their perceptions of the blogging community. We subjected the community scale to a component factor analysis employing varimax rotation with Kaiser Normalization. It yielded two distinct and reliable factors that we labeled Community Connection and Blogger Interaction. Community Connection (Cronbach $\alpha = .72$) included perceptions such as feeling connected with other bloggers and feeling a sense of community from blogging. The second dimension, Blogger Interaction (Cronbach $\alpha = .88$), included the activities of writing blogs, reading blogs, commenting on blogs, receiving comments on one's own blog, and interacting with bloggers outside of blogging (such as Facebook or Twitter). Participant scores for both dimensions ranged from 1.25 to 5.00 (M=4.09, SD=0.63).

After obtaining the raw data, we assessed the normalcy of the distributions by examining bar graphs of the variables' frequency distributions. Likeability, task accomplishment, moral, blogging interaction, and community connection did not display normal distributions. Because these variables appeared in each of the tested relationships, we conducted nonparametric statistical analyses to test the hypotheses, using the standard alpha level of .05, conducting two-tailed tests when calculating correlations.

H1 predicted a relationship between blogger's self-concept and content. As depicted in Table 1, a Mann-Whitney U test of independent samples compared the two content groups' scores across the six identity dimensions. Siegel (1956) explained that the Mann-Whitney U test for large samples serves as a good approximation to the randomization test and boasts a power efficiency of 100%. The analyses revealed no significant differences across the dimensions of likeability, task accomplishment, vulnerability, and moral. However, significant differences emerged between the groups on the identity dimensions of power and giftedness, with Family-Plus Bloggers reporting significantly more power and giftedness than Family Bloggers.

Table 1. Differences in Identity Scores
by Content Groups of Mommy Bloggers

Identity Dimensions	Family Bloggers Mean, SD	Family-Plus Bloggers Mean, SD	U	Alpha level
Likeability	36.04, 4.13	36.62, 3.95	36454	0.35
Power	24.20, 6.54	26.65, 6.12	29797	0.00*
Task Accomplishment	35.83, 4.43	36.38, 4.35	36512	0.29
Vulnerability	24.51, 6.58	23.27, 6.96	34594	0.13
Moral	26.10, 2.11	26.17, 2.32	38881	0.77
Giftedness	23.69, 4.88	24.82, 5.05	32651	0.02*

*Correlation is significant at the 0.05 level or below.

H2 predicted a relationship between the bloggers' identity and community. We calculated Spearman rank correlations to test the predicted relationships. As displayed in Table 2, five of the six identity dimensions correlated with sense of community; four dimensions correlated with blogger interaction. The higher the identity score on three dimensions (power, giftedness, task accomplishment), the higher both community scores (sense of community and blogger interaction). Conversely, the higher the community scores (sense of community and blogger interaction), the higher the three identity scores of power, giftedness, and task accomplishment.

Table 2. Spearman's Rho Correlation for Dimensions of Identity With Experience of Community

	Likea-bility	Power	Task Accom-plishment	Vul-nera-bility	Moral	Gifted-ness
Community Connection	*0.17	*0.07	*0.12	0.19	*0.10	*0.12
Blogger Interaction	0.06	*0.10	*0.05	*0.07	-0.02	*0.13

*Correlation is significant at the 0.05 level or below.

H3 predicted a relationship between content and experience of community. The Mann-Whitney U tests of independent samples revealed that the two content groups did not differ significantly across two of indicators of community ($U_{sense\ of\ community}$=39817, α=.47; $U_{blogger\ interaction}$=38146, α=.44). Given that the analyses failed to link community with content, we conducted a post hoc analysis testing our idea (Webb & Lee, 2011) that sense of community is linked to blogger interaction per se. We believed such an analysis was justified by numerous claims that the interactional nature of blogs facilitates the development of the blogosphere (see Dumova, 2012) as well as the results of our factor analysis in which sense of community emerged as an independent factor from blogger interaction. To that end, we calculated Spearman's rho rank correlations for bloggers' sense of community scores with their reports of blogger activities that comprise online interaction (reading, writing, and commenting). The results, displayed in Table 3, indicate that the stronger the sense of community, the more the blogger reported writing and commenting on blogs. Sense of community did not correlate with reports of simply reading blogs.

Table 3. Spearman's Rho Correlations for Blogging Interaction With Sense of Community

	Community Connection	
	Correlation Coefficient	Alpha Level
Read Blogs	0.04	0.18
Power	0.17	0.00
Giftedness	0.37	0.00

Note. Because these analyses were post hoc and no direction was predicted for the relationship, all correlations were two-tailed analyses.

Conclusions About the ICC Model

Based on the research reported here, Figure 11.2 is a further variation of the model originally presented in Webb and Lee (2011). It depicts the tested and confirmed relationships in the ICC Model of the mommy blogosphere.

H1 predicted a relationship between mommy bloggers' blog content and their identities. Results indicated that the two types of bloggers differed significantly in the identity dimensions of power and giftedness. Analyses revealed no significant differences between the two groups in their scores for likeability, moral, vulnerability, and task accomplishment. These limited but important differences may be the result of different foci, experiences, and acquired knowledge. The Family Blogger writes about her family, focusing inward (versus outward), perhaps interacting more with her family but less with the world outside her family, leading her to acquire more knowledge about her family but less knowledge about the outside world, and ultimately to feel less powerful—perhaps because her realm of influence is limited to her family. In contrast, the Family-Plus Blogger focuses both inward and outward, leading to increased interaction with the family as well as the outside world, and through that interaction learns more about the family and society. Thus, bloggers' focus, experience, and acquired knowledge may influence self-concept. Additionally, selected blogging topics may become increasingly salient to bloggers, and thus influence focus, realm of interactions, and knowledge of the world. A blogger potentially enacts identity on her blog, but in turn, identity could be influenced by the performance itself—creating a mutually influential relationship between the blog content and the blogger's identity.

H2 predicted a relationship between the bloggers' identity and sense of community. Becoming enmeshed in a community of people with similar interests who spend time engaging in the same activities may enhance the positive aspects of identity. This finding is consistent with Turkle's (1995) claim that identity can be influenced by online associations and communities. Similarly, people with positive identities may be more likely to blog confidently, telling the world about their achievements and challenges. The community of people following the same set of blogs may experience weak or strong ties, and both strong and weak ties can be beneficial in multiple ways, including offering social support to users (High & Solomon, 2011) and thus potentially influencing identity.

Specific aspects of identity seem to readily translate to experienced connectedness. For example, ranking higher in power may lead bloggers to write with a strong voice, attracting more readers and commenters, leading to a stronger sense of community. The less vulnerable and more confident she feels, the more she may be willing to share, and the more she shares, the more connected she may feel to others, because self-disclosure can lead to feeling close to others (Bane, Cornish, Erspamer, & Kampman, 2010). Then, the closer she feels to others, the stronger her sense of community.

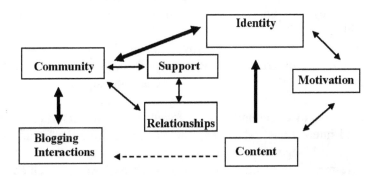

Figure 11.2. Tested and confirmed relationships in the ICC model of the mommy blogosphere. Bold lines indicate confirmed relationships. Solid lines indicate untested relationships. A dashed line indicates a tested relationship that remains unconfirmed.

Based on Webb & Lee, 2011. Published with the permission of Cambridge Scholars Publishing.

H3 revealed that community scores did not differ significantly by content groups. Instead, our findings lend credence to the claim that a sense of community may be independent of content and instead dependent on blogger interaction. Our post hoc analysis indicted that in our sample of mommy bloggers, sense of community was correlated with both writing and commenting on blogs, but not with merely reading blogs. Simply reading blogs fails to initiate or enact two-way interaction. Perhaps effective blogs, regardless of content, develop an interactive community that leads to both blogger interaction and a sense of community. Indeed, Stokes (2010) posited that using feedback mechanisms, such as the ability to comment, demonstrates the author's desire for approval, validation, and interpersonal communication. Scholars have argued that it is the interactional nature of blogs that makes them popular (Webb et al., 2012) and effective (Dumova, 2012).

Thus, interactional participation may be necessary to foster a sense of community. This finding is consistent with previous research documenting that talk among mommy bloggers prompts a sense of community (Tucker, 2009) as well as previous definitions of virtual communities as offering interactivity through comments and regular posting (Blanchard, 2006), and communities forming through talk between bloggers and their audiences (Tucker, 2009). This finding is also consistent with Kaye's observation that a "sense of community is developed through interactions with like-minded people" (2005, p. 76). An online blogging community is simply a group of people who exchange words and ideas through the mediation of a blog (Vickery, 2010). Therefore, having a stronger sense of community may prompt a blogger to interact more within the blog community.

Three findings combined may provide insight into the sense of community in the mommy blogosphere. The non-significant findings regarding a relationship between content and community, and the significant correlations between identity and sense of community, as well as between blog interaction and a sense of community, taken together may indicate that bloggers' identity and interactions, rather than their content, drive the formation and maintenance of blogger communities. Although many have argued that blog communities result from common interest in content, perhaps such content is a necessary but not sufficient condition for a virtual community. Instead, blogger interaction and identity may be the engines driving the formation and maintenance of blog communities.

Note

1. Further details concerning the sample are available upon request from the second author <LynneWebb320@cs.com>

References

Bane, C. M., Cornish, M., Erspamer, N., & Kampman, L. (2010). Self-disclosure through weblogs and perceptions of online and "real-life" friendships among female bloggers. *Cyberpsychology, Behavior, and Social Networking, 13,* 131–139.

Baym, N. K. (2006). Interpersonal life online. In L. A. Lievrouw & S. Livingstone (Eds.), *The handbook of new media: Updated student edition* (pp. 62–76). Thousand Oaks, CA: Sage.

Blanchard, A. (2006). Blogs as virtual communities: Identifying a sense of community in the Julie/Julia project. In L. Gurak, S. Antonijevic, L. Johnson, C. Ratliff, & J. Reyman (Eds.), *Into the blogosphere: Rhetoric, community, and culture and weblogs.* Minneapolis, MN: University of Minnesota. Retrieved February 27, 2012, from http://blog.lib.umn.edu/blogosphere/blogs_as_virtual.html

BlogHer network joins ranks of top women's networks online. (2007). *Media Report to Women, 35*(4), 1–3.

Bruns, A. (2005). *Gatewatching: Collaborative online news production.* New York, NY: Peter Lang.

Camahort, E. (2006, May 20). BlogHer '06 session discussion: *Mommyblogging is a radical act!* On day two. Retrieved February 27, 2012, from http://blogher.org/node/5563

Castells, M. (2004). *The Network society: A cross-cultural perspective.* Northampton, MA: Edward Elgar Publishing.

Davies, R. (2007). Networking for shy people: Building communities online. In *Making links: Fifteen visions of community* (pp. 71–78). London, England: Community Links.

Droge, C., Stanko, M. A., & Pollitte, W. A. (2010). Lead users and early adopters on the Web: The role of new technology product blogs. *Journal of Product Innovation Management, 27,* 66–82.

Dumova, T. (2012). Social interaction technologies and the future of blogging. In T. Dumova & R. Fiordo (Eds.), *Blogging in the global society: Cultural, political and geographic aspects* (pp. 249–274). Hershey, PA: IGI Global.

Friedman, M, & Calixte, S. (Eds.). (2009). *Mothering and blogging: The radical act of the mommyblog.* Toronto, Canada: Demeter Press.

Gurak, L. J., & Antonijevic, S. (2008). The psychology of blogging: You, me, and everyone in between. *American Behavioral Scientist, 52,* 60–68.

Haferkamp, N., & Krämer, N. (2008, May). *Entering the blogosphere: Motives for reading, writing, and commenting.* Paper presented at the annual meeting of the International Communication Association, Montreal, Canada. Abstract retrieved February 27, 2012, from http://research.allacademic.com/meta/p_mla_apa_research_citation/2/3/2/8/2/p232822_index.html

Hammond, L. (2010). Mommyblogging is a radical act. In J. F. Stitt & P. R. Powell (Eds.), *Mothers who deliver: Feminist interventions in public and interpersonal discourse* (pp. 77–98). Albany, NY: State University of New York Press.

Hatzipanagos, S., & Warburton, S. (Eds.). (2009). *Handbook of research on social software and developing community ontologies.* Hershey, PA: Information Science Reference.

Herring, S. C., & Paolillo, J. C. (2006). Gender and genre variation in weblogs. *Journal of Sociolinguistics, 10,* 439–459.

High, A., & Solomon, D. (2011). Locating computer-mediated social support within online communication environments. In K. B. Wright & L. M. Webb (Eds.), *Computer-mediated communication in personal relationships* (pp. 119–136). New York, NY: Peter Lang.

Hine, N. (2004, October). *Stories and blogs.* Paper presented at the annual meeting of the International Society for Augmentative & Alternative Communication, Natal, Brazil.

Humphrey, S. (2008). Grassroots creativity and community in new media environments: Yarn Harlot and the 4000 knitting Olympians. *Continuum: Journal of Media and Cultural Studies, 22,* 419–433.

Jost, K., & Hipolit, M. J. (2006). Blog explosion: Are blogs a passing fad or a lasting revolution? *CQ Researcher, 16,* 505–528.

Jung, Y., Vorderer, P., & Song, H. (2007, May). *Motivation and consequences of blogging in social life.* Paper presented at the annual meeting of the International Communication Association, San Francisco,

California. Abstract retrieved February 27, 2012, from http://research. allacademic.com/meta/p_mla_apa_research_citation/1/7/0/8/1/p170811_ index.html

Kaye, B. K. (2005). It's a blog, blog, blog, blog world: Users and uses of weblogs. *Atlantic Journal of Communication, 13*, 73–95.

Kleman, E. E. (2007). *Journaling for the world (Wide Web) to see: A proposed model of self-disclosure intimacy in blogs* (Doctoral dissertation). Kent State University, Kent, OH. Retrieved February 28, 2012, from http://www.personal.kent.edu/~ehollen2/research/Kleman %20(2008).pdf

Kline, D., & Burstein, D. (2005). *Blog! How the newest media revolution is changing politics, business, and culture.* New York, NY: CDS Books.

Kouper, I. (2010a). The pragmatics of peer advice in a LiveJournal community. *Language@Internet, 7*, 1–21. Retrieved February 27, 2012, from http://www.languageatinternet.org/articles/2010/2464/Kouper.pdf

Kouper, I. (2010b). Science blogs and public engagement with science: Practices, challenges, and opportunities. *Journal of Science Communication, 9*, 1–10.

Kwon, W., & Rudd, N. A. (2007). Effects of psychological and physical self-image on perceptions of salesperson performance and nonstore shopping intention. *Clothing and Textiles Research Journal, 25*, 207–229.

Lenhart, A., & Fox, S. (2006). *Bloggers: A portrait of the Internet's new storytellers.* Retrieved February 28, 2012, from http://www. pewtrusts.org/our_work_report_detail.aspx?id=21106

Li, J., & Chignell, M. (2010). Birds of a feather: How personality influences blog writing and reading. *International Journal of Human-Computer Studies, 68*, 589–602.

Lopez, L. K. (2009). The radical act of "mommy blogging": Redefining motherhood through the blogosphere. *New Media & Society, 11*, 729–747.

Mendelson, A. L., & Papacharissi, Z. (2011). Look at us: Collective narcissism in college student Facebook photo galleries. In Z. Papacharisssi (Ed.), *A networked self: Identity, community, and culture on social network sites* (pp. 251–273). New York, NY: Peter Lang.

Moravec, M. (Ed.). (2011). *Motherhood online: How online communities shape modern motherhood.* Newcastle-upon-Tyne, England: Cambridge Scholars Publishing.

Princeton Survey Research Associates. (2006). *Blogger callback survey: Data for July 5, 2005–February 17, 2006*, Retrieved February 28, 2012, from *Princeton Survey Research Associates International*: http://www.pewinternet.org/~/media/Files/Questionnaire/Old/PIP_Blogg ers_Topline 2006 pdf

Rocksinmydryer. (2008, August 1). *What do you think of the term "mommyblogger"?* [Blog post]. Retrieved February 27, 2012, from http://www.blogher.com/what-do-you-think-term-mommyblogger

Rodzvilla, J. (Ed.). (2002). *We've got blog: How weblogs are changing our culture.* Cambridge, MA: Perseus Books Group.

Rosenberg, S. (2009). *Say everything: How blogging began, what it's becoming, and why it matters.* New York, NY: Crown.

Shirkey, C. (2008). *Here comes everybody: The power of organizing without organizations.* New York, NY: Penguin.

Siegel, S. (1956). *Nonparametric statistics for the behavioral sciences.* New York, NY: McGraw-Hill.

Stake, J. E. (1994). Development and validation of the Six-Factor Self-Concept Scale for adults. *Educational and Psychological Measurement, 54*(1), 56–72.

Stake, J. E. (1998). *Six-factor self-concept scale.* Unpublished Scale, University of Missouri, St. Louis, MO.

Stefanone, M. A., & Jang, C. Y. (2007). Writing for friends and family: The interpersonal nature of blogs. *Journal of Computer-Mediated Communication, 13*, 123–140.

Stokes, C. (2010). "Get on my level": How Black American adolescent girls construct identity and negotiate sexuality on the Internet. In S. R. Mazzarella (Ed.), *Girl wide Web 2.0: Revisiting girls, the Internet, and the negotiation of identity* (pp. 45–67). New York, NY: Peter Lang.

Tucker, J. S. (2009). Foreward. In M. Friedman & S. Calixte (Eds.), *Mothering and blogging: The radical act of the mommyblog* (pp. 1–20). Toronto, Canada: Demeter Press.

Turkle, S. (1995). *Life on the screen: Identity in the age of the Internet.* New York, NY: Simon & Schuster.

Vickery, J. R. (2010). Blogrings as virtual communities for adolescent girls. In S. R. Mazzarella (Ed.), *Girl wide Web 2.0: Revisiting girls, the Internet, and the negotiation of identity* (pp. 183–200). New York, NY: Peter Lang.

Warburton, J. (2010). Me/her/Draco Malfoy. In S. R. Mazzarella (Ed.), *Girl wide Web 2.0: Revisiting girls, the Internet, and the negotiation of identity* (pp. 117–138). New York, NY: Peter Lang.

Webb, L. M., Fields, T. E., Boupha, S., & Stell, M. N. (2012). U. S. political blogs: What channel characteristics contribute to popularity? In T. Dumova & R. Fiordo (Eds.), *Blogging in the global society: Cultural, political, and geographic aspects* (pp. 179–199). Hershey, PA: IGI Global.

Webb, L. M., & Lee, B. S. (2011). Mommy blogs: The centrality of community in the performance of online maternity. In M. Moravec (Ed.), *Motherhood online: How online communities shape modern motherhood* (pp. 244–257). Newcastle upon Tyne, England: Cambridge Scholars Publishing.

Wei, L. (2009). Filter blogs vs. personal journals: Understanding the knowledge production gap on the Internet. *Journal of Computer-Mediated Communication, 14*, 532–558.

Woods, J. (2005). Digital influencers: Do business communicators dare overlook the power of blogs? *Communication World, 22*(1), 26–30.

Yanico, B. J., & Chih Lu, T. G. (2000). A psychometric evaluation of the six-factor self-concept scale in a sample of racial/ethnic minority women. *Educational and Psychological Measurement, 60*(1), 86–99.

CHAPTER TWELVE

Afterword:
A Remediation of Theory

Zizi Papacharissi

Historical context shapes how we interpret technologies. It informs uses, expectations, and rituals that are adjusted or re-invented. Technologies reorganize time and space conventions, presenting a *moving map of possibilities and interdiction* (de Certeau, 1984) or a *geography of the new* (Morley, 2007). Scholars are frequently drawn to the Cartesian dualism of time and space when speculating on the impact of technology, which invokes further dualisms revolving around mind and matter, the physical and artificial/non-physical, and an ensuing long succession of theoretically imposed binaries. Lewis Mumford (1934) employed the construct of the machine to explicate the interplay between technology and civilizations, via processes that were both socially shaped and shaping of technology. Carey (1989) told the story of the telegraph to explain the ability of technology to reconfigure time and space, thus restructuring conventional geographies and proposing new ones. Similarly, Marvin (1990) traced how the decoupling of time and space enabled by electricity and the telephone reorganized social relations, and Gitelman (2006) compared technologies of sound recording and digital recording to show how media become both subjects and instruments of history making. Historical context, that is, the connections between the past, the present, and our expectations of the future, defines how we use technology to repeat and to reinvent social routines, rituals, and habits. Historicizing suggests a linear chronology of technology, which implies a continuum of attempts at evolution.

This linearity positions technologies in sequential order, inviting recollections of the past in suggesting expectations of what might come next. Thus, expectations frequently inform the historical context against which we interpret outcomes. We may judge an outcome as unsatisfactory simply because it did not meet expectations derived from past experiences. At the same time, we are impressed by an outcome, because, we might say, it exceeded our expectations. We use standards or constants, which inform our interpretations of the world around us, but we also develop variable expectations based on our past experiences, So, a 140-character news update we receive on Twitter is judged as short, when compared to our expected length of news update delivered via a newspaper or television. Similarly, a text message is frequently perceived as brief, hurried, and filled with abbreviations that violate our expectations for thoughtful and polite expression. A letter, on the other hand, is valued because of the inconvenience it involves on the part of the letter-writer, who set time aside to compose a lengthier and presumably more thoughtful message. Linearity is suggested by expectations that come first, and technologies that follow.

Reverse Linearity

Perform, if you will, for the sake of argument, a mental exercise in reversing the linearity of technology. What if the journalistic ecology was defined by the norm of delivering around-the-clock 140-character-long concise updates of news events? Lengthier newspaper articles or features might then be judged as unnecessarily nuanced, verbose, and time consuming. What if our cultural habitus prized electronically mediated succinct expressions as efficient and essential to our everyday routines—a way for us to organize communication more efficiently? In this context, abbreviations employed in texting might be perceived as an exercise in mental acumen, and letter writing and reading might be interpreted as an imposition on a person's time.

In the same vein, what if the Internet preceded, rather than succeeded, television? Television might then be interpreted as a non-networked, more specialized, smaller version of the Internet. What if television came first, and print media followed? Print journalism might then be developed as an alternative to TV news, and not the other way round. What if the telephone arrived last, as a bare bones communicative option for those who wanted to commit to audio calls without the added visibility and performative multitasking that comes along with multimedia conferencing? These examples are not intended to be far-fetched. Future generations will not have grown up with newspapers, or television, and thus may not judge the Internet against the habitus of expectations that those who grew up with television as a novelty have. To these generations, the term *audience* might seem like a quaint expression used to describe folks engaged with a medium solely through the practice of listening. Likewise, the term *public* might confuse individuals who are accustomed to performing the self through perpetually public platforms such as Twitter.

Expectations shape and are shaped by the architectural design of technological platforms. Bolter and Grusin (2000) have described the process of *remediation* as one through which newer media achieve cultural singularity by mimicking, rivaling, and ultimately refashioning earlier media as they construct a unique identity of their own. Thus, the Internet attains its own place by borrowing from and extending elements of previous media, as did television, print media, radio, and other technologies. Remediation describes how expectations seep into technologies, shaping the uses to which we put them.

Similarly, I suggest that we permit the theoretical language we use to describe the present to borrow, apply, imitate, but also, ultimately, *reinvent* theory of the past. This deliberate practice may produce a theoretical vocabulary that singularly and irrevocably belongs to, and describes, the present. In this sense, processes of reverse linearity allow us to revisit, mimic, and even rival the past. In this form of theoretical existentialism, if you will, the present is an active ingredient in developing theory that will subsequently inform imaginings of the future.

The authors who contributed to this volume underscored the many ways through which new media remediate our everyday habits and routines, and in doing so, they invited readers to reconsider their scholarly assumptions, vocabulary, and methods. Social scientists interested in understanding the meaning of technology in everyday life must first challenge their own theoretical assumptions that frequently impose a linear analytical structure upon phenomena that are non-linear, convergent, and networked. The contributors to this volume understand the need to deviate from the linearity of mainstream paradigms, which follows the linearity of technological evolution, and which presents a theoretical canvas that both informs and restricts vernaculars we develop to understand the place of technology in our lives. Remediation of theory might consciously employ reverse or non-linearity as a way of enhancing the reliability and validity of theoretical observations.

Let me explain what I mean about linearity, and talk about how it disinvites the remediation of the theoretical assumptions, vocabulary and methods that are meaningful for research examining new media in the context of evolving uses of technologies. I mostly use the term to describe the heuristic tendency to describe, understand, and interpret our surrounding environments through the interconnected processes of cause and effect, or problem and solution. Thus, understanding a medium implies that we are able to determine the unique set of circumstances that enabled its existence. Further-more, interpreting the nature of a medium tends to be connected to establishing a typology of specific consequences, or effects, it might introduce.

In this manner, the paradigm of causality, which drives empirical research, is established and reinforced. Social scientific research subsequently accepts that in order to establish two phenomena as causally linked, correlation is a preliminary but not sufficient condition. Causality requires that correlation be present in a linear form; that is, one phenomenon must be shown to precede the other. And finally, it must be established that presence of the first phenomenon in the sequence is sufficient, on its own and in the absence of other contributory phenomena, to ensure that the second phenomenon will follow. What is identified as causality then is a direct line, of varying shape or form, which connects the two events together, upon a particular conjecture of space and time.

The paradigm of causality is meaningful because it is one way to organize our theoretical understandings and interpretations. More importantly, it permits the evaluation of possible causes and impact, and lays the foundation for the construction of hierarchies of concepts, constructs, perspectives, theories, and paradigms that are essential to organized scientific research. It leaves one asking further questions, however, because it presumes the possibility that phenomena may indeed be connected in a linear fashion. We presume, investigate and confirm causality on the basis of how we define it. The founding assumption is that processes that are connected must, at some point, have been correlated. These correlations take on linear shapes, and advance to various combinations of multilinearity, curvilinearity, and interconnection, but are never tangled; they have been unwoven and organized via the processes of heuristic investigation. Even when multimodal, they operate in a manner that assumes an order, sequence, and hierarchy.

I am certainly not suggesting that we abandon these practices. But I do suggest that we make a practice of becoming fluent in them and then forgetting about them on purpose, and building this step into the conventions of scholarly research.

We are afforded a unique opportunity to reinvent our research lingua franca via the architectures of networked technologies such as the Internet, which converge and collapse contexts, linearities, and perspectives. The reinvention of methods, vernaculars and assumptions might occur concurrently or in a linear fashion, or both. But the Internet, and the many platforms it presents, is not just a site of investigation but a vantage point, from which we are permitted to gaze upon old problems from a point of view that generates a different vista (Katz, 2010). Patterns and formations that are non-linearly related might present themselves through these different lenses, and all contributions to this volume on produsing theory might be read as such remediations and reversals of theoretical principle.

Bruns and Highfield describe an emerging ambient news environment, enabled by online media, that recasts both citizens and journalists as curators of information. In this environment, our conventional understandings of production and consumption of news are blended into nonlinear processes of *produsage*, which evolve beyond inadequate dichotomies that have long characterized how news is made, disseminated, circulated, and internalized. Citizens, citizen journalists, and journalists all become gatewatchers, engaged in produsage of journalism that involves many—discordant or coordinated—narratives, resulting not in a single, but a multitude of journalisms. Some of these journalisms are connected and others are parallel, moving on related paths that may or may not intersect.

Bolter invites us to consider how the theatricality afforded through online environments reinforces and challenges rituals of proceduralization that characterize late modernity. The emphasis on procedurality permits the understanding of how rituals of performance, upon which our social habitus is constructed, are augmented, reproduced, and reinvented online. Procedure and performance build on our previous understandings of social connection through self presentation in everyday life, but also evolve beyond that. The two processes, intertwined, help explain how individuals manage multiplied and multimodal selves that develop as we interact on a variety of social planes that associate procedures with the construction of performances of different aspects of the self. They are connected, but in a manner that evolves beyond the simplicity and heuristic order of causality.

The same elements of procedurality in performance are considered by Booth in *demediated* environments, where mediation is "obvious, ubiquitous, and (at the same time) effaced." Playspaces, enabled by interactive online games, present the ultimate demediation, as safe zones for performativity, representation, and immersion. Play in remediated, mediated, or demediated spaces is essential not only to presenting a character, but also in enabling a performance that may ultimately be aimed at disclosing an aspect of oneself. These performances can then be understood as disclosure through play; fundamentally "a public way to show private stuff" (Schechner, 2003, p. 265). The "as if" element of these performances, procedurally based as they may be, permits individuals to react to real interaction stimuli and simultaneously access fantasy and blocked material, transforming this all into a socially acceptable display or performance, and leading to what Schechner (2003) described as a form of *public dreaming*. In the collapsed context of converged environments, we might also understand behaviors as non-linearly connected and disconnected, publicly private and privately public, dreaming.

Heilferty examines the development of narratives of the self in mediated spaces reified through the practices of blogging. The research revealed the meaning of these narratives in addressing and managing not only uncertainty and stress, but also change against stasis, burdens versus gifts, and publicity *sans* privacy. The contributions of the study are important, but more significant, in my mind, is the flexible study design that evolves beyond conventional understandings that dichotomize online vs. other forms of support. Instead, these findings, and growing uses of mediated spaces for connection, expression, support and healing prompt us to consider integrating these into our habitus of mainstream parallel medical and health processes.

Hills, adopting a psychosocial or psychoanalytic approach to understanding digital fandom, suggests that demediated, remediated, or mediated spaces enable processes of connection and expression that engage us emotionally. I suggest that the procedures afforded for connection and expression online frequently engage the subconscious in a way that is emotional, and somewhat neglected in mainstream media research. And yet, the affective attachments we develop are connected to the social, cultural, political, or economic capital we accumulate, yielding primal emotive responses that anticipate (or, *premediate)* our performances.

Stern examines identity as a collaborative process, involving interpellation and performance. Newer media technologies amplify the multitude, immediacy, and hyper-sociality inherent in performances of the self, but in doing so, Stern argues, they focus identity processes on public identity production, rather than on public identity articulation. If we were to understand these processes as identity *produsage*, inclusive of collaboration, interpellation, and multi-performativity, we can begin to talk about remixed and remixable identities that are polysemically reified, so as to resonate with a multitude of publics and potential audiences without compromising one's own sense of self.

The storytelling project of the self, then, reflexive by nature and in search of a common linear connective thread, becomes saturated with opportunity for social interaction and publicity that emphasizes both the articulation of a self-story and the multiple potential interpretations thereof. The reflexivity and publicity extend beyond the story of the self, to the social roles the self takes on that both drive and restrict the potential selves accessible to the individual (Gergen, 1991). As Lee and Webb demonstrated, the mommy blogosphere represents a structure that not only supports but also renegotiates the boundaries of performances associated with family roles and motherhood.

In the same vein of expected and restored performances, Costa positions his theoretical renegotiation of established conventions. Costa's discussion of the bull as a discursive prosthesis that defines the boundaries of how we problematize the body, and as a result the digital outsourcing of sexual labor, is an exercise in the very process of theoretical remediation I have been describing. Operating in reverse, and thus, not building upon theory but rather, by deconstructing theory he presents a polyprismic contestation and reinterpretation of how we theorize scenes of racial politics and sexuality.

Also on biopolitics, but from a different perspective, Freedman considers how the technobiographic subject is re-assembled through biofeedback devices that afford a networked body. This way of looking at how technologies amplify and restrict human capabilities evolves beyond the metaphors of prosthesis or extension. It permits an understanding of integration, bypassing metaphors that view technology as separate attenuations of the human body. Technology is human, as are the politics of technological architectures, their design, and their alignment with or disruption to the human body. Still, evolving beyond the dichotomous approaches of technophobia and technophilia requires focusing on the in-between of the two binary positions.

Finally, Yep, Olzman, and Conkle examine specific narratives *prodused* via the "It Gets Better" project to understand how and why processes of progress are typically connected with practices of looking or *feeling backward.* Alternatively, practices of feeling forward are structured upon both the narratives of the past and the affective responses thereto. Were we to draw a parallel between feeling forward and thinking forward, we might consider how our theoretical understandings of problems past define, but also restrict, the lenses we use to view, feel, and think forward.

Learning (and) Forgetting

Education is founded upon honing our processes of remembering as a way to enable literacy, awareness, and critical thinking. As a result, we place emphasis on attention, and discourage distraction. We understand the two as opposite, when in fact, they both present different layers of concentration. Thus, we may tend to associate positive qualities with attention and negative consequences with distraction, when in fact, different forms of attention and distraction are prone to both. Similarly, we tend to evaluate our accumulation of knowledge by the ability to recall it, and having worked hard on this, a reluctant to forget it. When memory fails us, we regret not remembering, and devalue being able to forget. And yet, learning from the past but also being able to forget is frequently presented as a meaningful strategy of moving forward in folklore practices or formalized psychoanalysis. Memory is rightfully emphasized, because it facilitates learning and appreciation of knowledge. I suggest that rather than framing memory and forgetting as opposites, we conceive of them as parallel processes that at times connect, overlap, and diverge.

Our dominant educational heuristic prescribes logical progression and theoretical organization. Prime value is placed on discovering connection, describing it, explaining it, and predicting it. In the social sciences, we typically develop research and theory in a manner that reflects, reinforces, and perpetuates this linearity. We respond to new technologies by conducting research designed to address the policy concerns they generate (Dennis & Wartella, 1996). This is an established practice, and it is good practice, because it connects science to the public. Rather than abandoning this convention, we might become more aware of it and develop convergent, parallel, or divergent lines of inquiry that complement, extend, and yes, sometimes replace dominant approaches. Perhaps we might learn from artists, who become familiar with rules so that they are able to break them purposefully, to great effect. In doing so, Vincent van Gogh, Pablo Picasso, Salvador Dali, and many other artists each produced an early array of works defined by and upholding the contemporary artistic conventions, but then broke those rules in remediating artistic practices and presenting new paradigms structured around impressionism, cubism, and surrealism respectively.

Similar breaks with paradigms occur in the social sciences, and any number of socially oriented platforms facilitated by the Internet afford the different vantage point that may lead to the development of new paradigms and evolution of older ones. Availing ourselves of this new vantage point requires a few exercises in cognitive re-adjustment. First, we must become well-versed in learning and then forgetting. Emphasis should be placed in becoming fluent in our research conventions, so that we may we feel comfortable forgetting about them as we encounter new possibilities for studying and understanding social phenomena. Convention and disruption thus become interconnected parallel processes rather than linear opposites. Second, a new vantage point suggests that we understand attention and distraction as not opposites, but similar interconnected components of a process of cognitive concentration. Distraction is not inattention, but rather, attention to something else: the diversion of attention to an object of greater interest. Through parallel processes of attention and distraction we can learn and forget, as we reinvent our own habitus of research conventions. Finally, for Internet studies and the production of theory, the meta-heuristic implies that we allow the medium to define our choice of theory, method, and ultimately, perspective. Linear and non-linear, this form of thinking will serve the study of our everyday objects of attention and distraction.

References

Bolter, J. D., & Grusin, R. (2000). *Remediation: Understanding new media.* Cambridge, MA: MIT Press.

Carey, J. (1989). *Communication as culture.* New York, NY: Routledge.

de Certeau, M. (1984). *The Practice of everyday life.* Berkeley, CA: University of California Press.

Dennis, E., & Wartella, E. (1996). *American communication research: The remembered history.* New York, NY: Routledge.

Gergen, K. J. (1991). *The saturated self: Dilemmas of identity in contemporary life.* New York, NY: Basic Books.

Gitelman, L. (2006). *Always already new.* Cambridge, MA: MIT Press.

Katz, J. (2010, June). *Theory development and postconvergence: Challenges and opportunities.* Paper presented at the International Communication Association annual convention, Republic of Singapore.

Marvin, C. (1990). *When old technologies were new: Thinking about electric communication in the late nineteenth century.* New York, NY: Oxford University Press.

Morley, D. (2007). *Media, modernity and technology: The geography of the new.* London, England: Routledge.

Mumford, L. (1934). *Technics and civilization.* New York, NY: Harcourt, Brace & Company.

Schechner, R. (2003). *Performance theory* (Rev. ed.). London, England: Routledge.

Contributors

Jay David Bolter holds a PhD in Classics from the University of North Carolina and is currently the Wesley Chair of New Media at the Georgia Institute of Technology. He is the author of *Turing's Man: Western Culture in the Computer Age* (1984); *Writing Space: The Computer, Hypertext, and the History of Writing* (1991; second edition 2001); *Remediation* (1999), with Richard Grusin; and *Windows and Mirrors* (2003), with Diane Gromala. In addition to writing about new media, Bolter collaborates in the construction of new digital media forms. With Michael Joyce, he created Storyspace, a hypertext authoring system. With Blair MacIntyre and collaborators in the Augmented Environments Lab at Georgia Tech, he is exploring use of the Argon browser to dramatic experiences for games, art, entertainment, and informal education. His collaborations in AR experience design extend to Sweden, where he is working with Maria Engberg, Susan Kozel, and other researchers and designers at Malmö University and the Blekinge Institute of Technology.

Paul Booth is an assistant professor in the College of Communication at DePaul University. He is the author of *Digital Fandom: New Media Studies*. He earned his PhD in Communication and Rhetoric at Rensselaer Polytechnic Institute in Troy, NY. His research is focused on fandom, popular culture, media technology, and inter-media analysis.

Axel Bruns is an associate professor at the ARC Centre of Excellence for Creative Industries and Innovation (http://cci.edu.au/). He earned his PhD at the University of Queensland. He is the author of *Blogs, Wikipedia, Second Life and Beyond: From Production to* Produsage (2008) and *Gatewatching: Collaborative Online News Production* (2005), and the editor of *Uses of Blogs* with Joanne Jacobs (2006). His research blog is at http://snurb.info/, and he tweets at @snurb_dot_info. See http://mappingonlinepublics.net/ for his current social media research.

Allen Conkle is a graduate student at San Francisco State University. His interests include media and cultural studies, queer theory, and performance studies.

Diego Costa holds an MA in Cinema Studies from New York University. He is currently a PhD candidate in the Interdivisional Media Arts and Practice (iMap) program at the University of Southern California, where he teaches in the Gender Studies department. His research interests include queer theory, Lacanian psychoanalysis, childhood studies, digital humanities, and the relationship between new media technologies and sexual practice.

Eric Freedman is Professor and Dean of the James L. Knight School of Communication at Queens University of Charlotte, and earned his PhD from the School of Cinematic Arts at the University of Southern California. His research tackles several interrelated subjects that are included in the broad terrain of new technology, media access, and autobiographical discourses. He is currently writing on the industrial applications of game engines. Dr. Freedman is the author of *Transient Images: Personal Media in Public Frameworks* (2011).

Catherine McGeehin Heilferty, PhD, RN, is an assistant professor of nursing at Holy Family University in Philadelphia. She recently completed graduate study at Villanova University. Her research interests are an extension of her experiences over 20 years as a pediatric nurse caring for children and families in hospitals, home care and at the end of life. These research interests include family Internet use during life-threatening or chronic illness in childhood and its effect on the illness experience. In addition, she plans to examine the effects of uncertainty on quality of family life during childhood illness.

Tim Highfield is a research fellow at the ARC Centre of Excellence for Creative Industries and Innovation, Queensland University of Technology, and a sessional academic at Curtin University. He was awarded a PhD from Queensland University of Technology for a comparative study of political blogging in Australia and France. His current research examines how social media are used within public debates. His Web site is at http://timhighfield.net/

Matt Hills (PhD, University of Sussex) is a reader in media and cultural studies at Cardiff University. He is the author of five books: *Fan Cultures*

(2002), *The Pleasures of Horror* (2005), *How to Do Things with Cultural Theory* (2005), *Triumph of a Time Lord: Regenerating Doctor Who in the Twenty-first Century* (2010), and *Cultographies: Blade Runner* (2011). Matt has also published widely on cult media and fandom, and has book chapters forthcoming on topics such as cult stardom, the BBC TV series *Sherlock*, fan re-readings of the film *Inception*, and approaches to acafandom.

Brittney D. Lee (MA, University of Arkansas) is Arkansas Best Corporation's Marketing Coordinator with the MSI Group; she writes for Web sites, blogs, and other online venues. Her research examines gender and online communication.

Rebecca Ann Lind is an associate professor in the department of communication and associate dean of the college of Liberal Arts and Sciences at the University of Illinois at Chicago. A former broadcaster and print journalist, she earned her PhD at the University of Minnesota. Her research interests include race, gender, class and media; new media studies; media ethics; journalism; and media audiences. She recently published the third edition of *Race/Gender/Class/Media 3.0: Considering Diversity across Audiences, Content, and Producers* (2013).

Miranda Olzman is a second year MA student at San Francisco State University where she teaches fundamentals of oral communication. Most of her work has been in performance studies, but for her thesis she plans on delving into queer theory and fat studies. She also enjoys reading about critical communication pedagogy.

Zizi Papacharissi (PhD, University of Texas at Austin) is Professor and head of the communication department at the University of Illinois at Chicago and Editor of the *Journal of* Broadcasting *and Electronic Media*. Her work focuses on the social and political consequences of online media. She is author of three books and over 40 journal articles, book chapters, or reviews.

Shayla Thiel-Stern's research concerns the intersections of digital media, identity, youth culture, and gender. Her book, *Instant Identity: Adolescent Girls and the World of Instant Messaging*, was published in 2007. She is an assistant professor in the School of Journalism and Mass Communication at the University of Minnesota, and she earned a PhD from the University of Iowa.

Lynne M. Webb (PhD, University of Oregon) is a professor of communication, University of Arkansas. She has published two scholarly readers, including *Computer Mediated Communication in Personal Relationships*, and over 60 essays including work in *Journal of Applied Communication Research* and *Computers in Human Behavior*. Dr. Webb is known as an applied scholar and ground-breaking researcher in social media and in family communication who has published multiple theories, research reports, and pedagogical essays.

Gust A. Yep (Ph.D., University of Southern California) is a professor of communication studies, core graduate faculty of sexuality studies, and graduate faculty of the EdD program in educational leadership at San Francisco State University. His research, which has been published in numerous (inter)disciplinary journals and anthologies, focuses on communication at the intersections of culture, race, class, gender, sexuality, and health. He was an editor of the National Communication Association's Non-Serial Publication Program and he is currently serving as an associate editor of several communication and sexuality journals.

Index

General Editor: **Steve Jones**

Digital Formations is the best source for critical, well-written books about digital technologies and modern life. Books in the series break new ground by emphasizing multiple methodological and theoretical approaches to deeply probe the formation and reformation of lived experience as it is refracted through digital interaction. Each volume in **Digital Formations** pushes forward our understanding of the intersections, and corresponding implications, between digital technologies and everyday life. The series examines broad issues in realms such as digital culture, electronic commerce, law, politics and governance, gender, the Internet, race, art, health and medicine, and education. The series emphasizes critical studies in the context of emergent and existing digital technologies.

Other recent titles include:

Felicia Wu Song
  Virtual Communities: Bowling Alone, Online Together

Edited by Sharon Kleinman
  The Culture of Efficiency: Technology in Everyday Life

Edward Lee Lamoureux, Steven L. Baron, & Claire Stewart
  Intellectual Property Law and Interactive Media: Free for a Fee

Edited by Adrienne Russell & Nabil Echchaibi
  International Blogging: Identity, Politics and Networked Publics

Edited by Don Heider
  Living Virtually: Researching New Worlds

Edited by Judith Burnett, Peter Senker & Kathy Walker
  The Myths of Technology: Innovation and Inequality

Edited by Knut Lundby
  Digital Storytelling, Mediatized Stories: Self-representations in New Media

Theresa M. Senft
  Camgirls: Celebrity and Community in the Age of Social Networks

Edited by Chris Paterson & David Domingo
  Making Online News: The Ethnography of New Media Production

To order other books in this series please contact our Customer Service Department:
  (800) 770-LANG (within the US)
  (212) 647-7706 (outside the US)
  (212) 647-7707 FAX

To find out more about the series or browse a full list of titles, please visit our website:
  WWW.PETERLANG.COM